Erotic Games

EROTIC GAMES

Bringing Intimacy and Passion
Back Into Sex
and Relationships

Gerald Schoenewolf, Ph.D.

A Birch Lane Press Book
Published by Carol Publishing Group

A Birch Lane Press Book
Published by Carol Publishing Group
Birch Lane Press is a registered trademark of Carol Communications, Inc.
Editorial Offices: 600 Madison Avenue, New York, N.Y. 10022
Sales and Distribution Offices: 120 Enterprise Avenue, Secaucus, N.J. 07094
In Canada: Canadian Manda Group, One Atlantic Avenue, Suite 105, Toronto, Ontario   M6K 3E7
Queries regarding rights and permissions should be addressed to Carol Publishing Group, 600 Madison Avenue, New York, N.Y. 10022

Carol Publishing Group books are available at special discounts for bulk purchases, sales promotions, fund-raising, or educational purposes. Special editions can be created to specifications. For details, contact Special Sales Department, Carol Publishing Group, 120 Enterprise Avenue, Secaucus, N.J. 07094

Manufactured in the United States of America

10 9 8 7 6 5 4 3 2 1

Library of Congress Cataloging-in-Publication Data

Shoenewolf, Gerald.
    Erotic games : bringing intimacy and passion back into sex and relationships / by Gerald Shoenewolf.
        p.   cm.
    "A Birch Lane Press book."
    ISBN 1-55972-284-3 (hardcover)
    1. Sexual excitement. 2. Sex. 3. Sex in marriage.
4. Games for two. I. Title.
HQ18.S55   1995   94-44305
306.7—dc20   CIP

Contents

Preface

Today sexuality is often viewed more in terms of its possible hazards than its potential joys. In an age when sexually transmitted diseases have become all too fatal and complaints of sexual abuse fill the air, the positive aspects of sexuality and tenderness, including their restorative qualities, have all but been forgotten. We are deluged daily with news about the rising divorce rate and instances of marital incompatibility and violence. We are accosted by television talk shows that trot out an endless array of dysfunctional couples (not to mention triples and quadruples), demonstrating the full range of destructive "relating" that seems so prevalent nowadays. We see our friends and acquaintances struggle with their intimate relationships, sometimes choosing to abstain rather than to "take arms against a sea of trouble."

Erotic Games is, to my knowledge, the first organized book of therapeutic diversions. Whereas lovers of all varieties have played sexual games since ancient times, and some of those have been recorded, they are *not* therapeutic per se. On the other hand, certain exercises developed by sex therapists, although in fact therapeutic, are not really much fun. This book offers games that not only are both therapeutic and fun, but also are tailored to the various kinds of sexually blocked couples.

I must admit to having tried out many of them with my mate (Oh, what I have endured for the cause!), as well as to having prescribed them in my own analytic practice. Over the years I have found that such games, assigned as homework at the right time, can be a good way to help a couple break a par-

ticular impasse. Through trial and error, I gradually developed
a repertoire of games for use with different couples in various
situations.

It is my hope that these games may be as beneficial to others as they have been to my own patients—namely, in putting
sexuality back into a positive mode.

1

Introduction: Junk Sex vs. Loving Sex

"Doc, I don't understand women today," he said.

"Why not?"

"They're weird."

"What happened?"

"I was at my summer share this weekend and I went out with one of the women in my house, and we had sex—sort of."

"Sort of?"

"Right. See what you make of this, Doc."

"Go on."

"I really don't understand it, but maybe you can."

"I'm listening."

He paused to find the words, sighing and sitting back in his chair. He was a successful young man in his early thirties who had lived in Manhattan for several years, was buying a condo in the East Thirties, and had just broken up with another woman who had angrily accused him of having a fear of intimacy. He had rented a room in a summer house in the Hamptons, hoping to wash out the bitter taste from this last relationship with a lot of salt air, ocean, and, most important, more noncommittal sex.

"Listen to this, Doc. I go to a disco with a bunch of people from the house, and I end up dancing with this one woman. After a while I ask her if she wants to get something to eat. She smiles at me and then says, 'Yes.' Then when we're sitting at the restaurant she says, 'So, I guess you want to get laid?' 'Sure,' I said. 'That would be nice.' We go back to the house and she sits in my lap and—get this—she starts masturbating and sucking her thumb and tells me to say dirty things in her ear. So I start saying dirty things—you know, like 'You dirty slut,' and things like that—and she really gets off on it; and I'm thinking this is just foreplay, that later I'm going to get laid, as she had promised. But then she comes and says, 'That was nice, thanks,' and waltzes out of the room and I'm left holding the—holding—myself."

He shakes his head and looks at me with his handsome brown eyes. His voice has a slight whine in it, but his eyes have a mischievous glint, and I can see he is not really that perturbed. It is as though he is saying, "Yes, it was a bit strange, and I got teased—but wasn't it exciting?" "What do you think, Doc?" He raises and lowers his eyebrows three times. "Is that weird, or what?"

This patient's experience is typical of those of many on today's dating scene. Yet, just a few decades ago it would have been atypical. In former times, our society had a much different attitude toward dating, sex, and marriage. Indeed, prior to the early 1960s, popular magazines still debated about whether it was all right for people to have sex before marriage—and, in most of the societies of the world, premarital sex is still forbidden. In such societies there are clearly delineated rules to follow with regard to dating, sex, and marriage. You may agree or disagree with the rules, but you are not confused about your role, sexuality, or identity as they relate to those rules. Similarly, in former times, sexuality was less complicated: There were neither sex therapists to tell us about our various sexual disorders, nor social workers to remind us of the various kinds of sexual abuse. Sex just *happened*—and it was either good, or not so good. In former times there was no AIDS.

Today people enjoy a sexual freedom that perhaps no other society has ever enjoyed: There are scarcely any rules, as long as the sex is between consenting adults. However, while having too many rules may be stifling, *no* rules can be baffling. What would have been seen as perverse and indulgent in the past is now viewed as diverse and creative. What was viewed with forbidden joy is now seen with trepidation, something infested with the ever-present specter of AIDS and other sex-

ually transmitted diseases. What was once naive and sentimental is now often complex and clinical, surrounded by anx-
ieties pertaining to harassment or rape. In the Hamptons, on
college campuses, in marital beds, and in any other place
where lovers meet, there is often an atmosphere of distrust
between the genders.

The New Sexuality of today is also likely to be hurried,
since this is a generation that seems to have put relationships
on the back burner while emphasizing careers. We eat, work,
and sleep on the run. We have been raised in a culture of junk
bonds and junk food, and what we practice may aptly be
described as junk sex. It is junk sex because it is sex on the
run, sex in avoidance of the hazards, sex attempting always to
be correct, sex of convenience, sex *sans merci*.

Indeed, sex therapists across the country report that the
most common complaint among couples today is lack of interest. Many men and women are having shallow, infrequent, or
no sex whatsoever, simply because they do not care to. They
truly live lives of quiet desperation, and often their underlying
mood is apathy. Because of the hazards of sex and their own
internal resistances to it, they have given up on it. Their sexuality remains only in the fantasy sphere.

Like junk food, junk sex merely satisfies an appetite while
leaving deeper (dietary/emotional) needs unmet. It offers
immediate gratification without responsibility. Ironically, the
very responsibility that is avoided is what gives sex its most
profound meaning.

Should we return to another time? No, not necessarily. To
return to an earlier era is to return to childhood again; it cannot be done. But perhaps we can retain our present sophisti-
cation while incorporating some of the values that have
worked in previous eras and in other societies—including
mutual respect, trust, and caring between men and women; a
spirit of sexual fun; and a greater commitment to sexuality as
a part of an ongoing relationship.

As goes sex, so goes the relationship—that is the premise of
this book. If two would-be lovers are blocked in the sexual
sphere, they will be blocked in other spheres as well. If they
are unable to achieve authentic sexual bonding, they will be

unable to truly bond with themselves or with others. Their social, professional, and parental lives will all suffer. Most important, their relationship with themselves will also suffer.

Hence, if there is any one avenue that can be crucial to restoring love and tenderness to a marriage, as well as authenticity and meaning to one's life, it is the sexual avenue. Other methods may serve the same purpose, but sex seems to be the most direct route.

There are some truisms about sexuality that bear repeating. Sexuality, quite obviously, is much more than the mere act of copulation. It is nature's way of ensuring the continuation of the human species, and, as such, it is the grandest act of all, with many layers of meaning. Any sex is better than no sex, but at its most mature and deepest level, sex brings about the most profound kind of intimacy and bonding. It forces us to confront and resolve our most stubborn barriers to intimacy— barriers to both ourselves and others. If we can truly relate to even one other person, we can then relate to ourselves and to all others also.

Each act of sex brings us closer to that other person. The more mature the sex, the more authentically intimate and tender that sex will be. In that sense, each act of sex is a triumph of love and life (Eros) over hate and death (Thanatos). Freud, who saw life as a struggle between these two forces, believed that the main cause of human misery was the sexual repression necessitated by modern civilization. The antidote to civilization and its discontents is not *more* sex but *better* sex. The more genuinely we engage in sex, the more our bodies and minds will come alive and be freed from the chains of negative thinking. The more we live and love, the longer we will refuse to die.

As I wrote previously in *Sexual Animosity Between Men and Women*, every form of human pathology, whether psychological, organic, or social, is in some way associated with sexual frustration. Thus, neurotic defenses, narcissistic grandiosity, borderline rage, addictions, and psychosis—to name but a few—are all products, in part, of the frustration of Eros and the stirring up of Thanatos. The neurotic distorts the sexual experience with guilt, as when an obsessive demands neat and orderly sex and thus makes the experience ritualistic; the grandiose narcissist needs to be superior and cannot allow for the authentic vacillations of human sexuality; the addict is wedded to an addiction and can have only impulsive and shallow sexual experiences; and the psychotic withdraws completely from sex and people alike. In each instance, sexuality becomes blocked, and the other becomes an object to be distanced, exploited, distrusted, or hated.

Likewise, social and organic diseases often have a psychological component that is somehow related to frustrated sexuality. Prisons hold individuals whose sexual drives have become twisted and whose aggressive drives have become stuck in the mode of trying to destroy the environment that has failed to meet their needs. Maladies such as cancer, heart disease, and stroke may also be connected in part to chronic sexual frustration. I have observed, in my own practice, that those who suffer from cancer invariably are emotionally and sexually alienated—that is, they are the living dead—and generally harbor deep sexual conflicts.

Another way of putting this is that the more energy an individual expends on defense (resisting intimacy with others), the less one will become truly actualized as a human being and in harmony with others and with nature. Certainly, it is necessary to be defended in the everyday world. The trouble is that most people are much too defensive when they should not be, and not defensive enough when they should be. They are out of tune. Sexuality is nature's way of helping humans retune themselves, to work through unnecessary resistances and the underlying pent-up feelings.

Melanie Klein, the child psychoanalyst, noted that during intercourse an individual's frustration is soothed and aggression is lowered:

> Libidinal satisfaction diminishes his aggressiveness and with it his anxiety. In addition, the pleasure he gets from such satisfaction seems in itself to allay his fear of being destroyed by his own destructive impulses and by his objects [inner demons], and to militate against . . . his fear of losing his capacity to achieve libidinal satisfaction. Libidinal satisfaction, as an expression of Eros, reinforces his belief in his helpful imagos and diminishes the dangers which threaten him. . . .

How important is human physical contact? It is so important that at critical periods people will die without it. René Spitz conducted a study of ninety-one infants in a foundling home who had been separated from their mothers. These infants were fed by a succession of nurses and given only perfunctory physical contact. Thirty-four of them died by the second year, while others went on to develop numerous symptoms. All of the infants showed manifestations of severe separation anxiety; then angry crying and clinging (trying to control or destroy the frustrating environment); then anaclitic depression (aggression turned against the self); then motor retardation; then marasmus (withdrawal into apathy); and finally—all too often—death. Others have replicated this study.

Similar drastic responses to the frustration of Eros can be observed in adults, as when a husband or wife suddenly dies and the spouse dies a short time later, without there having been any normal precipitating illness; or as when a scorned lover goes on a destructive spree of gambling, drug abuse, or promiscuous sex or kills his mate or himself.

It is in our first relationship with our mother that we experience the prototype of love and tenderness. Love is an offshoot of gratitude. The original gratitude the infant feels toward the gratifying mother and her gratifying breast is both sexual and emotional. A mother's voluntary giving of love to her infant, who is too helpless to control whether she does so or not, is the first act of love. If the mother gives to the infant in this way, the infant will express gratitude and passion toward the mother, and the mother will experience a mutuality of tenderness and love, and a bond will be formed. Both will feel loved and appreciated, emotionally as well as physically. If a mother, due to her own emotional blocks, is unable to set this first example, the infant will develop blocks to intimacy.

The seeds of resistance to intimacy are planted in early childhood, watered throughout adolescence and young adulthood, and become fully grown sexual blocks in adult life. These blocks manifest themselves in a myriad of defensive postures, such as angry defiance or martyrdom or anxious submissiveness or appeasement or perfectionism—all of which are attempts to control both ourselves and others. As long as we are controlling and manipulating people, we cannot allow genuine intimacy to develop. It must develop voluntarily and cannot be manipulated or controlled.

Hence, mature sexuality is free of control or manipulation. It is a voluntary sharing of affection between two consenting adults. Just as a mother voluntarily shares her breast with her infant, so a lover voluntarily shares his or her sexuality. In this voluntary sharing between two consenting adults, a mutuality occurs which becomes the framework for mature loving. You cannot will this process to happen by determining to be more loving. Rather, it comes about through the acceptance of everything about ourselves and our lovers—our hopes, our fears, our beauty, our ugliness, our loves, our hates, and in particular everything we would like to disown and disavow in both ourselves and others.

If, as I contend, sexuality is the key to restoring vitality and tenderness to marriage, to self-actualization and to harmony with nature, why doesn't everybody just have more sex? Of course, it is not that simple. It is not so easy for people who have been blocked since early childhood (or who have developed blocks following a traumatic situation in adulthood) to suddenly become unblocked and be capable of achieving deep and gratifying sexual play leading to love. It requires far more than the intention to do so. Usually it entails help from others—a therapist, a doctor, an understanding friend, a lover, a spouse.

The games in this book are designed for use as an aid to becoming unblocked sexually and emotionally. They can be used in conjunction with therapy or with the assistance of a doctor. They can be used by therapeutically sophisticated couples for a self-directed form of sexual therapy. Or they can simply be used for fun.

These games differ from the usual games that lovers play in that they are designed to be therapeutic. The usual games that lovers play are compulsive. They are enactments of rituals that bring about a temporary gratification, but not the deeper joy of mature, tender merger with another being. Promiscuous individuals will have multiple lovers and sometimes daily sexual experiences—but their sexuality will have a numb, addictive quality to it, with weak or nonexistent orgasms, and it will never quite satisfy. Sadomasochistic people will repeat the same perverse rituals of domination and degradation again and again, deriving "highs" from it but never really achieving the liberation they seek. Narcissistic people will not be able to come out of themselves enough to really be with another person, so their sexuality will be a repeating pattern of sexual exploitation. Fetishists will act out rituals in which sexual energy is displaced onto an object (such as a shoe or a glove) or a part of a woman's anatomy (the forearms or breasts or feet), and the act will become alienated and devoid of its primary meaning.

These kinds of games do not lead to any resolution because they are linked to fixations in childhood and are in actuality defenses against intimacy. These defenses were developed on the heels of traumatic events. A three-year-old girl, for example, will discover that she has a vagina and that it feels good to rub her fingers (and other things) against it. She runs to her father and says, "Daddy, look at my vagina!" If the father has a block against sexuality, he may reply, "Don't do that. That's dirty." Or he may become sexually exploitative. Or he may just make a face. Anything that he does, and does repeatedly, will then affect the girl's sexual development.

Such a little girl, when she grows to adulthood, may form a narcissistic pride about her vagina and her sexuality as a defense against her father's and other men's scorn or exploitation, and her sexual relations as an adult will become blocked. Her father's original scorn or exploitation of something that she thought was so normal and natural remains as a scar in her psyche. She will always be a bit fearful of men, anticipating their scorn of her vagina and her sexuality; and she may even suffer from a form of sexual disorder. Her sexuality will be compulsive and defensive, a ritualized act serving to defend against scorn or exploitation—which to her translates to a repudiation of her whole self. If a person has a fixation, he or she retains the unconscious primitive fears that engendered the fixation.

The games in this book are designed to lead to the resolution of sexual and emotional blocks. In a sense, they are an elaboration of the kinds of exercises devised by William Masters and Virginia Johnson as an adjunct to the sex-therapy program they introduced in *Human Sexual Response*. However, their exercises are aimed at the resolutions of such sexual symptoms as premature or retarded ejaculation, partial or full impotency, partial or full frigidity, sexual phobias, and the like. The games offered here are intended to reach deeper into the unconscious recesses of character formation; the aim is not only to relieve symptoms but to resolve characterological attitudes.

The main difference between the games in this book and the sex games that people normally play is that these games are not compulsive rituals but rather conscious, deliberate enactments with a special purpose. They are similar to the games people normally play in that they address the same fixations that compulsive rituals address, but they do so in a way that is designed to resolve (rather than maintain) them.

These games do not stay in the same stage as do the common sexual games, but rather evolve through two stages. The first stage comprises games that address a particular symptom with their typical underlying fixations. These games, which encompass most of the chapters in this book, include everything from "Games for Bored Couples" to "Games for Unattracted Couples." The games are to be repeated as often as necessary until a couple feel that they have accomplished their purposes, which we assume have to do with revitalizing their interest in sex, their sexual passion, and their commitment and love for each other.

The second stage is made up of "Games to Restore Tenderness." Again, these games should be repeated as often as necessary until tenderness is revived. Once tenderness is rekindled in their marriage, couples find that they become more adjusted and tolerant and easygoing with other people, developing a "fellow-feeling" toward humanity that might have been lacking before, or which they might have had but not with much emotional intensity. (Note that the intense feelings that individuals have for religious or social causes are not to be confused with fellow-feeling, which consists of acceptance of all people, regardless of their views.)

The last chapter describes a harmonious couple—a couple in harmony with themselves, each other, and the world. It is a harmony that the ancient Chinese sages referred to as yin and yang, a harmony that consists of being (as the ancients called it) "deeply rooted and firmly stemmed."

Let me close with a few disclaimers. This book is not just for young or beautiful people, but for couples of all ages and physical types. Sexual passion can be had by anyone who wants it. Age can diminish the sexual drive but does not eliminate it; only organic or psychological barriers can do that. Physical appearance can be used as an excuse to avoid sexuality, but behind every physical molehill is a mountain of anger, jealousy, or fear. And, although this book is addressed to heterosexual married couples, it can also be used by homosexual or unmarried couples who have a long-term, committed relationship.

Most therapists will tell you that the prettiest patients can (and sometimes do) develop erotic transferences toward the ugliest therapists. It so happens that sexual attraction, at its deepest level, does not depend on physical appearance—which only serves as an initial spur. At bottom, sexuality depends on one's capacity to give and receive gratitude and admiration with grace, to accept and share profound vulnerabilities without fleeing from them, and to tolerate anger and rage without giving in to the impulse to flee, appease, guilt-trip, or get revenge.

Having said that, let me add that this book is not a sexual panacea. If you use it according to instructions, and are willing to persevere even when it seems as if nothing is happening (even when, in the beginning, it seems silly, pointless, hopeless, upsetting, or disgusting to continue), then you may well achieve results. These are sure to be gradual, even though there may be an initial upsurge of fascination and joy—as happens in the beginning of most any endeavor.

Also, these games may provoke painful feelings or memories, and even reawaken traumas; hence they may be upsetting and even at times unnerving. I therefore strongly advise that they be used in conjunction with psychotherapy. If you use them without a psychotherapist's supervision, do so just for fun, realizing that you are at your own risk. Like any other tool, they can be misused as well as used.

Finally, these games presume the practice of "safe sex"— sex in which the couples have an exclusive relationship or are using proper precautions to avoid sexual diseases.

Bon voyage!

2

Games for Bored Couples

Bored couples are not really bored. They are experiencing a kind of suspended animation. Boredom is a state of mind that occurs when wishes, fantasies, and feelings are being repressed because, if admitted into consciousness, they would cause anxiety.

Generally only one partner is feeling boredom, but on occasion both are. One of my patients, a man in his late thirties, complained to me of being bored by his marriage: "My wife is a very boring lady. She's a complainer. All she does is complain, complain, complain. But if I say anything to her about her constant complaining, she accuses me of not being empathic enough. She just wants to complain but never wants to really examine herself. She can never be there for me. Even when we have sex, which isn't very often, I feel she's just sort of taking a break between complaints."

This patient's boredom was a defense against both the anger he felt toward his wife for constantly complaining and shutting him out and the taboo wishes and fantasies he harbored of a sexual or violent nature. His wife was doing to him what she did to every other man (creating distance and desexualizing the relationship), and the patient was doing to her what he did to every woman (subtly rejecting her emotionally, and depriving her by withholding his anger).

The resolution to such an impasse is to communicate what is being repressed. However, this is not so easy. To simply tell my patient to express his anger or verbalize his taboo fantasies of strangling her or of having angry sex with her would do no good. He would think of a thousand reasons for not doing so.

His resistance to expressing such things, formed in relation to his mother, has become a deep character trait. Likewise, his constant complainer of a wife has a history of using whining to manipulate men into feeling sorry for her even while repressing her own anger and feelings of low self-worth. Her troubled history began with her relationship with her own father, a passive man who never took her seriously. And while consciously she would maintain that she's all for communication and that it's her husband who holds back, in actuality she too is resistant to genuine communication.

It can take months—even years—of therapy for such couples to break through these resistances. However, through erotic games, the wishes, fantasies, and feelings will often be prodded loose rather more swiftly as a by-product of play. The spontaneity associated with sexual play undermines rigid character defenses and hastens confrontations that would otherwise continue to fester.

Following are five games to be used by bored couples. Games 1 and 2, "Seduction Surprise (by the Wife)" and "Seduction Surprise (by the Husband)," are variations of the same game. These would be excellent starting games for the husband and wife described above. Which game should be played first depends upon who is "boring" and who is "bored." In the case just discussed, it would probably be the husband who would be the game activist, since he would feel most sexually deprived and therefore eager to try something new. In cases where it is the wife who is feeling frustrated sexually (or otherwise), and therefore most bored, she would be the game activist.

Both Game 1 and Game 2 are designed to address the sexual fantasies underneath the surface boredom, and in doing so to jar loose the barriers (suppressed thoughts) to intimacy and pleasure. Game 4, "Deserted Island," is another game that addresses the sexual fantasies in order to loosen rigid character resistances. However, it also adds elements of novelty and entrapment as aids for truly bored couples. Games 3 and 5, "Who Cares?" and "Sexual Confession," are directed at barriers to intimacy rather than at sexual fantasies. The first utilizes a paradoxical approach that therapists call "joining the resistance," encouraging bored couples to express—and even exaggerate—their feelings of boredom, as well as other feelings that constitute the barrier. The second uses nakedness and seduction as a kind of mesmerizing, double-dose truth serum.

Couples may experiment with these games and settle on whichever seem most applicable. Or they may devise variations of their own. "Repeat as often as necessary"—for a week, a month, a season—until you *do* experience changes in your relationship. As you replay these games, boredom will be replaced by an array of other feelings—anxiety, jealousy, fear, lust, anger, sadness, love—the gamut of feelings that humans are heir to. Embrace all of these feelings as though they were lost children—which in a sense they are. When passion (intense feelings about one another) is restored, you are ready for chapter 14, "Games to Restore Tenderness."

Game 1: Seduction Surprise (by the Wife)

Players: Boring husband and bored wife.
Activist: Wife, without husband's cooperation or knowledge.
Setting: Home.
Aim: Prod husband out of his defensive posture and rekindle sexual passion and emotional involvement.
Game Plan: The husband comes home from work to find the house in darkness. When he flicks the light switch, nothing happens—the fuse has been disconnected. There is just enough light from flickering candles so that he can make his way to the living room.

There in the middle of the living room, lying languorously on the floor, is his grinning wife. She has made a nest in the middle of the floor by piling up comforters, pillows, or rugs, giving the room the appearance of a harem boudoir. Perhaps she has even hung some diaphanous material, something akin to pink chiffon, from a makeshift rafter—and she lies within this mysterious, tentlike shroud. On each side of the nest are candles or lanterns (Japanese lanterns are nice), and the smell of incense wafts through the air. From somewhere comes the sound of exotic music on strange instruments.

His wife is clad in something soft and sexy—a silk nightgown or robe, harem pants, gauzy slip. Through this outfit he can see her panties. They could be red panties, because red provokes passion as well as anger—though what's best is whatever color the wife knows most turns him on (sometimes white, virginal panties really do the trick, and sometimes black works best). She smiles seductively at him.

"What's going on?" he may ask.

"Whatever you like," she will answer.

If he is an easy case, he may drop his briefcase and enter the nest forthwith. If he is a more difficult case, he will resist this seduction in various ways. He may say, "This is silly," or, "I'm too tired for this," or, "What about dinner? I'm starved." Or he may say, "Where'd you get this material? How much did all of this cost?" He may even get surly and insulting (his anger defending against the anxiety that this Seduction Surprise has aroused) and blurt out, "You look stupid lying there with that idiotic grin."

No matter what he says or how long or in how many ways he resists, the wife is to persist with her seduction: "I know it's silly, but come join me anyway," she says; or, "I know you're starving, but come join me for a while anyway"; or, "Yes, it cost a few dollars, but come join me for a while and you'll see it was worth it"; or, "You think I look idiotic? Come join me and you'll see how truly idiotic I can look!" Repeat these phrases as many times as needed, until the husband surrenders.

For extremely difficult cases, the wife may need to do something more physical. She may, for instance, have some champagne (or his favorite other drink) ready and can try to loosen him up with that. She may offer a massage for his tired shoulders. She may perform a belly dance or striptease. Or, she may crawl up to him on her knees, unzip his pants, and take a hands-on approach to seduction. Each case will be different, and in each case the wife must rely on her own understanding of what her husband's particular weakness is. If he's obsessed with the stock market, she may suddenly show miraculous knowledge of that day's activity. If he loves baseball, she may suddenly reverse her usual aversion to it, don a baseball cap, and begin a conversation about that day's game, the statistics of the opposing pitchers, the need for a new manager, and so on.

It is crucial for the wife to never, never give up. She must regard this as a battle—which it is! Her husband has maintained his defenses against spontaneity and intimacy for a good reason (let's say, for example, he had a very intrusive mother), and he may fight almost to the death to protect himself against vulnerability. The wife must therefore be prepared to fight this battle until he finally gives in, never taking any of his refusals or insults personally, never insulting him back or in any way losing her temper, but always sticking firmly and seductively to the game plan.

Once he gives in to the seduction, the rest of the game is easy. Having lured him out of his defensive posture (of being boring), and herself out of her own defensive posture (of being bored or frustrated), there will most likely be a newfound passion for one another, as well as a newfound interest in sex.

The game can then be repeated and varied—or, a couple may simply begin having more sex without the game. And more sex will at least temporarily bring about a stronger bond that will enable the couple to begin communicating about things they have long kept pent up. "All you do is talk about the stock market," she may say, "and that makes me feel angry and rejected." He may reply, "I wish you'd take more of an interest in my work and stop being so demanding." At the very least, a dialogue is started—and, once this dialogue has begun, the couple is ready for chapter 14, "Games to Restore Tenderness."

Warning: Soon after the flame of intimacy is rekindled, there is a tendency to retreat to the original defensive postures—and both partners must be wary of that. It will be an ongoing struggle against giving in to this resistance, one that may last for weeks, months, even years. Defensive postures are just another kind of addiction—but, like all other addictions, they are hard to break. It takes a valiant effort to do so.

Game 2: Seduction Surprise (by the Husband)

Players: Bored husband and boring wife.
Activist: Husband, without wife's cooperation or knowledge.
Setting: Home.

Aim: Prod wife out of her defensive posture and rekindle sexual passion and emotional involvement.

Game Plan: The wife comes home from work (or wherever) and finds a note on the front door. "Hello, my dearest wife. You have been elected queen for a night. Prepare yourself for the surprise party of your life and times!" She enters to find that the lights are low, the scent of incense is in the air, and the strains of soft, exotic music (or the romantic songs of her favorite crooner) fill the room. When she enters the dining room, she finds the table set with their finest china, napkins, and silverware, candles burning, and her favorite flowers in a vase at the center of the table. The aroma of steaming oysters (or her favorite food) comes from the kitchen. A bottle of champagne in a bucket of ice sits at the corner of the table.

"Good evening, my dear," the husband says, popping out of the kitchen, dressed in a tuxedo. "Here, let me help you with that." He takes her purse, her briefcase, her packages— whatever she is carrying.

"What's going on?" she may ask.

"Shhhhh!" He holds a finger to his lips. "Sit down." She sits. "Good," he says. He then places an imitation (but attractive) crown on her head. "I hereby crown you queen for a night."

"This is silly," she may say; or, "I'm too tired for games"; or, "What are you up to?"

"Shhhhh!" Would my lady care for a glass of champagne now, or after her bath?"

"My bath?"

"Yes, your bath."

If she is a difficult case, she will attempt to get out of this in some way: "This is all very nice, but I'm not in the mood," or, "You look silly in that tuxedo." She may even become obnoxious and insulting. The husband must persevere and not take anything she says personally. And he may cleverly (for once) use his knowledge of her to persuade her.

"Here—let me massage your temples," he may say (or, "Let me massage your shoulders"). He can try a trade-off: "Humor me tonight and I'll do something for you some other night." If all else fails, he can explain that he is playing a therapeutic game recommended by a psychotherapist, and that if she wants to improve their marriage it would be beneficial for her to play along—at least for one night. Eventually, if the husband perseveres, he will wear down her stubbornness, and she will give in. Then he can proceed.

"Come, my lady," he says. He escorts her into the bathroom, where he has prepared a bubble bath. He undresses her and helps her into the tub, then fetches her a glass of champagne. As she sips the champagne, he gently bathes her, taking special care as he washes her private parts. Looking caringly into her eyes, he then invites an unexpected pleasure: "Would you like your hair shampooed, my lady?"

"Yes, please."

He shampoos and rinses her hair, after which he intones, "Now, just lie back a while and relax. I have to check the dinner." He returns in a few minutes and helps her out of the tub, dries her with a towel, blow-dries her hair, and then hands her a package containing a new silk robe, saying, "I think my lady will be more comfortable in this." He helps her on with the robe, standing behind her, kissing the back of her neck as he does so. "Sorry, my lady, I forgot myself for a moment." He puts her crown back on and leads her by the hand to the dining table and properly seats her. "More champagne?"

"Yes, please."

He pours another glass and brings out the oysters and salad. As they eat, he asks, "How was your day, my lady?"

"My day?"

"Yes. Please tell me everything."

"You don't want to know."

"Yes—I want to know everything."

The husband makes a point of listening to everything his wife says, in a way he never has. If the wife starts to complain, and he finds himself becoming bored, he is to say to himself (silently), "She's complaining because she doesn't feel appreciated deep down," and he is to lean forward and pay even closer attention to what she is saying. "The more I give to her," he tells himself, "the more she will be able to give to me." He leans forward and listens as closely as possible while he sensuously sucks on oysters.

When they have finished, he says, "Would you care for coffee or tea?" He brings the beverage on a tray, adding,

"Don't worry about the dishes. I'll do them later." (he eventually carries them into the kitchen, rinses them in the sink, and returns.) "And now, for the final treat of your queendom—if my lady please." He takes her arm and leads her into the bedroom, which is also lighted by candles. He has decorated the bed in an unusual way; it is covered by satin sheets and pillowcases, and colored paper streamers hang from the light fixture (or from the bedposts).

In the bedroom, he begins to make love to her in the way that he knows she likes best. Usually wives have told their husbands over and over what they like, but bored (i.e., angry) husbands usually do not listen, since their wives' sexual preferences sound like more complaints or demands.

All except the most hardened and resistant wives will respond to such an evening. While the way to a man's heart may be through his genitals, the way to a woman's heart is through her head. Catering to her wishes, listening to her, and taking her seriously (even when she is being demanding) are what generally serve to seduce her. By the end of the evening, she will have been drawn out of her defensive posture of complaining, resisting, or being "too busy." Her sexual passion will be revived, along with the beginnings of a new tenderness. In playing out this game, the husband will also have to relinquish his defensive posture of boredom or irritation—and his passion will also be rekindled.

Once they have revived their sexual passion and formed a new erotic bond, the couple will feel more secure about revealing their honest feelings about one another. The husband may now say, "You know, I was afraid to tell you before that I feel really irritated and shut out by your complaining and your distractedness [etc.]." "And I feel upset that you never listen to me and are always watching football," the wife may say.

This game may be repeated as needed—if not in deed then in spirit. It should serve to shake things up and cause the couple to begin to look at themselves in a new way. They may also try the other games in this and other sections, as applicable. When ready, the participants may go on to the games in chapter 14, and play all of the games in that section.

Game 3: Who Cares?

Players: Bored husband and bored wife.

Activists: Both husband and wife take part in creating this game.

Setting: A room with two plain chairs facing one another.

Aim: To force a husband and wife out of their boredom by giving it expression while adding an erotic element. This method is referred to by psychotherapists as "joining the resistance." In this case, both the husband and wife, ironically, join each other's resistance.

Game Plan: The husband and wife sit facing one another. The husband begins the game by saying, "I don't feel like having sex."

The wife replies, "I don't feel like having sex."

"It's boring," the husband says.

"Who cares?" the wife retorts.

They may repeat these or similar phrases as often as necessary. At first they may find themselves snickering, not really "feeling" them: The words may sound artificial. But if you are truly a bored couple, these phrases (or something similar to them) should eventually feel right.

Then the husband removes an article of his clothing—a belt, a hat, a ring—and says, "This is boring, but I'll do it anyway."

The wife also removes an article of clothing and says, "Yes, this is really boring, but I'll go along with it."

They proceed to remove all their clothing, repeating the same or similar phrases. Then the husband reaches over and caresses the wife's breasts. "How does that feel?" he asks.

"It doesn't matter," she replies. She may even yawn.

"Do you want me to continue?" he asks.

"I don't care."

"Do you want me to stop?"

"I don't care."

As he continues, she fondles his genitals. "How does that feel?"

"It doesn't matter."

"Do you want me to continue?"

"I don't care."

"Do you want me to stop?"

"I don't care."

They continue the foreplay, trying new things. As they give vent to their feelings of boredom while engaging in this kind of sex play, they will begin to feel less and less bored. Then, in the second phase of the game, as their passion begins to rekindle, they should follow each statement of boredom with the very next thought that enters their mind.

"Does this excite you?" the husband asks the wife.

"Not at all," the wife says. Then, her next thought: "Well, actually, I kind of like it."

And the wife asks the husband, "How does that feel?"

"Who cares?" he may say. Then: "*I* care!"

And the wife may say, "I'm bored"—and then, "I don't understand why I'm getting so excited."

And the husband may say, "It doesn't matter"—and then, "I feel afraid of losing control."

And the wife may say, "Who cares?"—and then, "I think I've been feeling upset with you for years and holding on to that."

And the husband may say, "Boring!"—and then, "I'm so angry at you for distancing me all the time. I think you need a good fucking!"

This game may or may not lead to actual sexual intercourse the first time it's played. Instead, the first few times might result only in "seducing" buried feelings and bringing them to the surface—feelings such as lust, fear, anger, or jealousy. A rule of therapy is that if an individual is afraid to feel negative emotions, he or she won't be able to feel positive ones, either. So if we suppress any feelings, we end up suppressing them all. Once suppression and repression are lifted, there may be a temporary euphoria of liberation—followed by anxiety and then a resurfacing of the "negative" emotions we have been holding on to, denying, or projecting onto others.

This third game may be repeated as often as needed, until it leads to sexual intercourse, passion, and more-honest communication. It may also be used in combination with other games.

Game 4: Deserted Island

Players: Bored husband and wife; travel agent.

Activists: All three players consent to and participate in the game.

Setting: Deserted island in the middle of a lake, or cabin in the woods.

Aim: To throw husband and wife together into an intriguing (nonboring) situation from which there is no escape, and force them to relate in a new way.

Game Plan: The travel agent (a friend or relative) blindfolds the bored husband and wife and takes them to a deserted place. The agent may take them by boat to an island in the middle of a lake, or by car to a cabin in the middle of the woods, and leave them there. In either case, they will be stranded in an unfamiliar place for a weekend. The island or cabin will basically offer only food, water, and shelter—no modern amenities, such as television or laser disks. If either the husband or the wife should want to escape, he or she will have to swim or walk a long distance, even to a back road.

Before the travel agent leaves the couple, he or she asks the husband and wife each to hand over a previously agreed-on sum of money (for example, $100), then says the following: "I'm leaving now and will return in two days. During that time you may only make love if you do so in a way you have never done before, and each time before you make love you must each say to the other, 'We're all alone here,' and then express the next thought that comes into your head. There's a telephone in the cabin, so you can call me if you want me to come get you. If you call me before the weekend is over, or if you haven't made love in three new ways, you forfeit the money."

The agent repeats these instructions several times, so that the effect borders on the hypnotic, and asks both husband and wife separately to repeat the instructions verbatim. Only when they are able to do so does the agent leave.

The couple find themselves thrown off-balance in many ways. They are lost physically, stuck in a strange environment without any idea of where it is located. They also are in a no-win situation psychologically, because if they do not have sex in three new ways, they'll lose not only their money but also the game—yet if they do have that sex, they will be forced to do something they have been studiously avoiding precisely because it is anxiety provoking. They are literally deserted, with nobody to turn to—except each other.

Having them say "We're all alone here" before they have sex, and then say the next thought, helps them get in touch with what they're feeling, and is likely to bring up stuff they have been suppressing. The wife may say, "We're all alone here," and then, "Oh, God—now I have to depend on *you!*" The husband may say, "We're all alone here," and then, "Oh, God—now I'll have to put up with your nagging!"

During the course of the weekend they will be forced in this way to confront many issues that have lain dormant. In addition, by having sex in new ways, they will be forced to work through sexual phobias that have served as defenses against intimacy, and this will in turn force them to confront even more issues. Hence the channels of intimacy, passion, and communication will be opened.

If they cannot "succeed" with the game the first time, and call the agent to pick them up, they can try the routine again until they get it. Such weekends may be repeated as needed, and may be combined with other games in this book.

Game 5: Sexual Confession

Players: Interrogator and confessor.
Activists: Both husband and wife.
Setting: Home or hotel.
Aim: Using nakedness and erotic touch as a "truth serum."
Game Plan: The husband and wife lie or sit facing one another on a bed or sofa. They are naked. They take turns being the interrogator and the confessor.

The designated interrogator reaches out, fondles the genitals of the confessor, and asks, "Do you like that?"

"Yes, I like it."

"Do you want me to continue?"

"Yes, please continue."

"Do you want me to stop?"

"No, don't stop."

"Tell me one truthful thing about myself that you've never told me."

"I can't think of anything."

The interrogator stops fondling the confessor.

"Please don't stop."

"Then will you tell me one truthful thing about myself that you've never told me?"

"Yes."

The interrogator starts fondling the confessor again.

"Tell me."

"I feel afraid of you."

"That's better. When do you feel afraid of me?"

"I feel afraid of you right now. You've got your hands on my genitals."

"That's true."

"And I also feel afraid of you at other times."

"What other times?"

"I'm not sure."

The interrogator stops fondling the confessor.

"Don't stop."

"Then talk."

"I will."

"What other times?"

"I'll tell you. I'll tell you. Please don't stop. I feel afraid when you're drinking."

"That's better." The interrogator continues fondling the confessor until the whole truth is told. Then the two switch roles.

This game can last (intermittently, at least) for an hour, a day, a week, or a month, and constantly lead to new truths. By being truthful in the context of erotic play, the couple is enabled to open up in an enjoyable way. This provides positive reinforcement to the difficult task of breaking through barriers. Husbands and wives will be surprised at the things both their partners and they come out with. Of course, there may be some pain to deal with—unexpected truths that are difficult to hear and cope with. Then the couple must stop the game and deal with that pain, perhaps even with the help of a therapist.

Eventually (as with the other games in this section), boredom will dissipate while intimacy and passion return. Again, when the two are ready, they may move on to chapter 14.

3

Games for Passive-Aggressive Couples

Prior to the 1950s, therapists most frequently met with couples in which the husband was too aggressive and the wife too passive. Nowadays, perhaps due to changes in social values, we more frequently encounter couples among whom the reverse is true. Pierre Mornell, a clinical psychologist, wrote a book about this latter syndrome, giving it the humorous title *Passive Men and Wild Women.*

A patient I have treated for a while is involved in the latter-type marriage. Her husband, a devout Quaker, appears to be the perfect mate. In many ways he is very attentive to her: He cooks wonderful meals, does windows, is handy around the house, and never loses his temper. He believes he is a spiritual and concerned person. Twice a day he meditates, and often he goes away on meditation retreats. Yet—despite all this—my patient continually feels furious toward him, and even finds herself saying sarcastic things to him in front of people at parties, causing them to wonder (sometimes aloud) how come this nice guy puts up with such a monstrous mate. And every other month, she explodes, throws whatever she can grab at him, and pummels him with her fists.

"I don't understand it," she moans to me. "He seems to bring out the monster in me. He's so good, I can't stand it. He keeps wanting me to do meditation with him, implying that if I just calmed down, things would be alright. He thinks everything's my fault—that I'm hysterical, and that's why we have problems. If I yell at him, he just looks at me with a sad expression, like he's thinking, 'Why, God, have I been bur-

25

dened with such a witch?' When I complain that he isn't interested in sex, he says maybe he would be if I weren't so sarcastic to him. He always passes the buck back to me. I can't
stand him! I want to strangle him! I want him to disappear! I
hate him! He disgusts me!"

What happens in such a case is that the husband is dis-
owning his own aggression and projecting it onto the wife: "I
don't hate her. I'm not a hateful person. It is she who is the
hateful person." However, he passive-aggressively provokes
hateful feelings in her by frustrating her. This is all the more
galling because he is doing that while appearing to be nice.
The passive is an individual—and, as we have said, it can be a
woman as well as a man—who has a rigid ideal image that
must be lived up to at all costs. This person must not harbor
any hateful thoughts and must always be right and good. The
passive husband in our example was convinced that he felt
only love for his wife, and that if only she would calm down,
everything would be dandy.

This man could not grasp for a moment that he had an
unconscious need to infuriate his wife, so as to feel right and
superior—since she was always the person who was "losing
her cool." He had developed this way of defending in connection with an obsessive-compulsive mother, who had wanted to
organize and control every aspect of his life and know everything he was thinking and feeling. To survive, he retreated into
a world of passivity and meditation, keeping his thoughts and
feelings locked up where she could not get to them (and
therefore to him). Now that same mode of operation had been
displaced onto his wife.

Meanwhile, the wife had experienced a passive father, who
gave her anything she wanted materially, but was completely
absent emotionally. She always had the impression that her
father, a politician, married and had children only because it
was politically expedient to do so. Just as her own mother had
ranted and raved at this passive father, so now she too became
enraged at her passive husband. The more monstrous she felt,
the more she hated both herself and him, and the more she
would resort to name-calling, guilt-tripping, and temper
tantrums.

When husband and wife are locked in such a struggle, it is
difficult to get either of them to see what they are doing. Since

in this case the impasse centered on the wife's sexual frustration, I devised some games for them. The wife was too angry at the husband to want to seduce him (as in Game 1), so I had her do both "Role Reversal" (Game 4) and "Psycho Surprise" (Game 5). These games served to surprise and upset the husband, and thereby take him out of his passive-complacent mode. They got into an argument in which both admitted their frustrations, he for the first time acknowledging that he was furious at her because he found her to be contemptuous toward him: "If you treated me with respect, I'd have more sexual desire for you!" She begrudgingly changed, and their sexual relations improved.

Game 1: The Master and the Maid

Players: Master (passive husband) and maid (aggressive wife).

Activist: Wife.

Setting: Home.

Aim: The wife seduces the husband in a deliberately provocative way designed to provoke both his sexuality and his active, rather than passive, aggression.

Game Plan: The husband comes home from work (or from wherever he has been during the day) to find his wife wearing a maid's uniform with a short black skirt and white apron.

"What's with the uniform?" he may ask.

"It's a maid's uniform," she replies, mysteriously.

He goes into the bedroom to change into something more comfortable and then sits in his easy chair to watch the evening news. As he does so, his wife crosses in front of him and stoops down, with her rear toward him, to open the bottom drawer of a desk and straighten out its contents. Lo and behold, she is not wearing any panties!

The husband, true to his passive-aggressive character, will at first pretend he does not see what he sees.

The wife struts past him, smiling mysteriously, then returns to the desk again to dust off the top. Naturally, she must stoop to do this, and it is also of course necessary for her to wiggle her rear as it protrudes toward her husband. Her naked behind flexes this way and that, and now it is only a few feet from his face. He can smell a new brand of perfume she has apparently dabbed onto her secret region and can hear her humming something softly under her breath.

The wife continues to cross smilingly before him and to stoop provocatively in front of him, wiggling and swaying and dipping and squirming while fooling with the furniture and fixtures, until he cannot help but ask, "What are you doing?"

"Oh, just looking for something I once lost," she demurely replies.

"For something you once lost? I see."

"You see? What do you see?" She wiggles her rear some more.

At this point he will begin to feel both aroused and frustrated. He may respond by jumping up right then and rushing forth to take her from behind. Or he may get angry and snap at her, "Why don't you put on some pants? That's disgusting. You look like a whore." Or he may walk out of the room to avoid this seduction, which arouses feelings he has long strived to avoid and does not want to deal with.

The main difficulty for the wife is to overcome her accumulation of hurt pride. Her husband's passive-aggression and sexual frustration tactics may have left her bitter, and at the moment she might rather strangle than charm him. But she must transcend this impulse and channel it into her mischievous seduction. She must also not allow his initial avoidance or insulting behavior to derail her efforts, but must carry on, no matter what. Even if the seduction does not work on the first night or the second, it may just click four nights later.

Once he takes the bait, the wife should encourage him to make love to her aggressively. "Do it to me—do it, do it!" she may beg. "Yes, yes, yes—that's the way I like it." And during the act itself, she should encourage him to verbalize his anger at her. "I'll bet you really hate me sometimes, don't you?"

"That's right, you bitch, I do!" he may say.

"How much do you hate me?"

"A lot!"

"A lot?"

"Yes, a lot."

"Show me."

"I *am* showing you."

"Show me more."

"You brazen slut!"

"Yes!"

"You really are a brazen slut, aren't you?"

If it succeeds, this game serves two purposes at the same time: It gets the aggressive wife out of the aggressive, persecutory mode and into a more seductive, receptive mode; and it nudges the husband out of his passive-aggressive mode and into a more erotic, directly aggressive mode. In accomplishing this, the game also diminishes the frustrations of both husband and wife and facilitates a more honest and open discussion of both their sexuality and their relationship in general.

Game 2: Headache

Players: Passive wife and aggressive husband.
Activist: Husband, without wife's knowledge or cooperation.
Setting: Bedroom.
Aim: Resolve the defensive posture of the wife by "killing with kindness," and change the husband by having him give up his tendency to guilt-trip or threaten.
Game Plan: This game is for the passive-aggressive couple in which the wife is the passive and the husband is the aggressive, and their sexual relations are epitomized by the "Not tonight, dear, I've got a headache" syndrome. Couples who get stuck in this syndrome are invariably locked in a particular kind of impasse. In this fix, the wife appears to be a kind, giving person who takes care of the husband and children—and, indeed, often treats her husband as if he were one of the kids. But when it comes to adult sexuality, she avoids it. The husband likes being pampered by the wife and feels guilty about wanting more sex. His response to being sexually frustrated is to beg, guilt-trip, and threaten his wife: "Please, I need it!" he may say—or, "I don't know why I put up with a wife like you"—or, "Maybe I'll have an affair."

In this game, he surprises his wife one night by dispensing with his usual woe-is-me demeanor and adopting her own caretaker mode. To signal this change, he has donned a white bow tie, which he wears along with his pajamas. He slides into bed with her, smiling cheerfully, and asks, "So how are you tonight?"

"I'm okay. Why?" She does not at first look away from the television show she is watching.

"I just wondered. May I get you some juice?"

"No, thank you. What's gotten into you?" She finally glances at him, and her eyes rest first on his cheerful smile, and then on the bow tie.

"What's that?"

"It's a tie."

"I know, but why?"

"How's your head tonight?" he demurs.

"I have a slight headache, as usual."

"Would you care for a scalp massage?"

"A scalp massage?"

"That's right."

"I think I must be hearing things."

"No, you're not hearing things."

"You're just trying to soften me up so I'll have sex."

"I can understand your thinking that. Lie back. Relax."

He proceeds to give her a long, tender scalp massage. He must really give her a good, caring one in order for this exercise to work. It can't be a quick, perfunctory job. (If he does not know how beforehand, he can consult a "how-to" book.) When he finishes, he asks, "There, how do you feel now?"

"Wonderful."

"What can I do for you now?"

"I don't believe you."

"What don't you believe?"

"Why are you doing this?"

"Because I love you. What can I do now?"

"I can't think of anything. I'm too shocked."

"Well, if you do, let me know. I'm at your disposal."

"And you don't want sex?"

"I want it, but I know you have a headache and you're not in the mood. I understand."

"You do?"

"Yes. I mean, I love you and I love making love to you, but I don't want to make love to you unless you really want me. I don't want you to do it out of duty or because I browbeat you into it. All these years I've just been thinking about myself, about my own selfish sexual needs, and I haven't been thinking about you. So I thought that for once in my life I'd think about you."

"I can't believe my ears."

"Believe it. It's true."

"So you really don't expect sex?"

"No. I want it, but I don't expect it. When you feel ready to give to me in that way, I'll appreciate it from the bottom of my heart and loins. Until then, I'll just be patient."

"You know, that tie is kind of cute." She reaches out to touch the bow.

"Don't," he says, pulling back.

"Why not?"

"You can't remove the tie until we have sex."

This is the way the conversation might go, but it will of course vary according to the couple. (In some cases it may be the husband who gets headaches, and the wife will wear some white article of clothing, such as a nurse's or nun's hat—white symbolizing purity.)

If the active spouse can manage to pull off this abrupt change in attitude with sincerity, it can have amazing results. The wife may throw herself into his arms right away, disarmed by this new attitude and the mysterious, provocative bow tie. The fact of the tie, plus his statement that she could remove it only when they have sex, will immediately put their sexual relationship on a different plane: Instead of the whining complainer for whom she must do her duty, he becomes a charming, teasing challenge. And he will in fact be mirroring her own mode, offering care while withholding sex.

In more difficult cases, it may take a while to break down the wife's resistance. The husband must be prepared to wait as long as necessary, pouring on the kindness and bearing her skepticism and anger with grace. It is a game of wills, and he must funnel his own anger into this constructive battle, killing (defeating) her with kindness. Eventually, if he persists, she will surrender.

Once she does, a new balance will have been accomplished. She will have given herself to him not out of guilt or duty, but because at last she really wanted to. And he will have taken perhaps his first step in learning the value of constructive charm.

Game 3: Nude Hamlet

Players: Passive spouse (audience), aggressive spouse (actor), and dummy.

Activist: Aggressive spouse, without the knowledge or cooperation of mate.

Setting: Living room with makeshift stage or room with real stage.

Aim: To shock passive into awareness and allow aggressive to discharge rage.

Game Plan: This game is a take-off on the play within a play from Shakespeare's *Hamlet* in which he notes that "The play's the thing, wherein I'll catch the conscience of the king."

The aggressive spouse announces to the passive spouse one night after dinner, "Darling, I have a little surprise. I've made up a little play just for you. You like theater, don't you, darling?"

The aggressive spouse turns the lights down and prepares to act out a scene from their sex life. On the stage is a sofa or bed on which lies a life-size "dummy" (or doll) which will be the surrogate for the passive spouse. The aggressive spouse, having stripped down, enters the scene naked, and lies beside the dummy. (Let's generally refer to the passive-spouse "dummy" as *the dummy*, and to the aggressive spouse as simply *the spouse*.)

"Darling, let's make love," the spouse proposes.

"Not tonight—I'm not in the mood," the dummy replies.

"You're never in the mood," the spouse shoots back.

The dummy is silent.

"Talk to me."

The dummy looks at the spouse but remains silent.

"Talk to me."

The dummy shakes its head.

"Don't shake your head at me!"

The dummy yawns, holding a hand over its mouth.

"I hate you when you act like this!"

The dummy simulates rolling back its eyes.

"I hate you, I hate you, I hate you!" screams the spouse, who begins kicking and hitting and biting and scratching the dummy. The dummy lies back passively. "You make me feel like a monster sometimes. You make me feel like hitting you and hurting you until you do something!"

When the spouse lies exhaustedly crying, the dummy says, "It makes me feel so superior and good to see that you're a monster and I'm not. It makes me feel so good to know I'm completely innocent and have no hatred inside me whatsoever, while you are full of hate. But I will love you anyway, despite your faults, and just try to be patient until you see the light."

By this time the real passive spouse will have at least begun to protest. "This is a ridiculous play!" The active spouse will then invite the passive spouse to the stage to play the dummy's role, but will require him or her to undress first. They will do the scene again—but this time as the passive spouse would like to play it. When they do play out the new version, the active spouse will begin to embrace and kiss the passive spouse—a move which will lead to new and unexpected reactions and feelings on the part of both.

The passive spouse, in playing the scene differently, has the opportunity to do what every writer does—re-create life in one's own image. In doing so, that person unwittingly begins to see the relationship in a new way and to try new approaches and responses.

This couple definitely will find that their sexual relations improve as they replay this script. Also, the play will leave an indelible impression that will require much further discussion—if not right at that time, then at some point in the near future. The scene can be repeated again and again and each time elicit new reactions and feelings, stimulating a resurgence of sexual passion and more-honest communication.

Game 4: Role Reversal

Players: Husband and wife.
Activists: Both.
Setting: Bedroom.
Aim: Jolt each other out of their passive and aggressive defensive postures by mimicking each another.
Game Plan: Husband and wife lie virtually naked in bed. The husband wears an article of the wife's clothing—could be a hat, a blouse, panties, or shoes. The wife wears an article of the husband's clothing—hat, pipe, jockey shorts, T-shirt. They play out their continuing sexual conflict, but each impersonates the other, deliberately aping and exaggerating the other's behavior.

If the couple were my aggressive female patient and passive husband, about whom I wrote in the beginning of this chapter, the wife might sit up in bed with her eyes closed and pretend to meditate. The husband might hit her (lightly) with his fists and complain that she is such a nerd. The dialogue might go something like this:

"Can't you see I'm meditating?" (Wife impersonating the husband.)

"Oh, you're such a nerd." (Husband impersonating the wife.)

"Leave me alone. I'm meditating."

"I don't know why I married such a nerd. Why couldn't I have married a real man?"

"Quiet! I'm trying to achieve cosmic consciousness! Stop trying to do dirty, earthly things to my body while I'm looking at universal harmony."

"When I think about the kind of man I always dreamed of marrying and the one I ended up with, I want to puke!"

"Oh, God, save me from this witch!"

The dialogue may go on like this for a while. At first the couple may be inhibited about playing each other, unable to get out of themselves. But if they keep at it, they will find themselves enjoying exaggerating each other's defensive postures, and they will also benefit from seeing their own behavior exaggerated by the other. What often happens is the dialogue becomes more and more exaggerated until one of them exclaims in surprise or anger, "I don't act like that!" This leads to a discussion.

Another level of this game occurs when the couple practices role reversal during intercourse. If the man is usually on top, ejaculates prematurely, and tends never to look in the woman's eyes, then the woman now plays out that role—getting on top, refusing to look in the husband's eyes, pretending to reach orgasm quickly, and then dismounting and turning away (to meditate). And if the woman is usually on the bottom, takes a long time to reach orgasm, and does not gaze into the man's eyes, the man now plays her that way. And so on. This again leads to such exclamations as, "I don't act like that!"

By mirroring each other, each one is forced to take a more objective look at himself or herself. Often they get angry at each other's exaggerated portrayals, but sometimes they break into laughter. Laughter helps to release the tension and the pent-up frustration and anger that have accumulated during the course of the marriage. And by seeing their partner enact their own means of avoiding sexual intimacy, they often begin to stop avoiding and (for example) actually look one another in the eyes during sex, even if that proves at first somewhat uncomfortable. Looking at each other eventually becomes the lesser of two evils, because they come to view *not* looking at each other as even more shameful.

Family therapists have long used role reversal as a technique in couples therapy. It is even more effective in the bedroom, where eroticism gives it a stronger impact. Using eroticism as a tool and role reversal as a technique, a couple will eventually take a new look at their usual mode of operation, begin to discuss new options, and come to enjoy a new sexual passion and comraderie. The aggressive partner becomes less aggressive, demanding, or controlling, and the passive partner less passive, appeasing, or controlling.

Game 5: Psycho Surprise

Players: Passive spouse and "psycho" spouse.

Activist: Aggressive spouse, without the knowledge or cooperation of passive spouse.

Setting: Bathroom.

Aim: To enact the very fears that the passive spouse unconsciously harbors, and thereby shock him or her out of the passive mode.

Game Plan: This game is a more radical maneuver and thus entails more risk than most others. It is a take-off on the famous shower scene from Hitchcock's *Psycho*. While the passive spouse is having his or her morning or evening shower, the aggressive spouse sneaks into the bathroom, wielding a toy knife—one of those knives made of soft rubber, plastic, or foam, which bends at impact. The aggressive spouse throws aside the shower curtain or door, glares at the showering spouse, and laughs maniacally.

"Die! Die! Die!" the "psycho" yells as he or she brings the rubber knife down onto the body of the passive spouse.

After the stunned victim has screamed with fright and stands gaping at the "psycho," the latter stops suddenly and looks at the former coldly and calmly for about thirty seconds, tosses away the knife, grabs the victim, and plants a firm, passionate kiss on his or her mouth.

What happens next may vary. The passive spouse may continue to stand gaping, and the aggressive one may walk out of the bathroom—and only later will they talk about it. Or the passive one may become angry and reprimand the aggressive spouse. Or the episode may lead to a wild lovemaking scene in the bathroom. In any event, something out of the ordinary will happen.

If all goes according to plan, the passive spouse will be provoked out of his or her passivity, and will express the real anger and hate that underpins the passive-defensive posture. Since that person already views the spouse as a monster, this scene sets off—even exaggerates—this fantasy, forcing the passive to recognize and verbalize what was formerly secretly thought. It also provides the "psycho" spouse with a way of expressing her or his anger at the passive spouse in a constructive way (a way that has a humorous undertone and leads to resolution) rather than through destructive sarcasm, guilt-tripping, or temper tantrums.

This game should result, for the first time, in honest communication about what is going on between them. It should also lead to a revival of sexual interest and passion.

Variation: A variation of "Psycho Surprise" has the "psycho" spouse slide into bed some night wearing a monster face or witch mask—and nothing else. The "psycho" laughs maniacally and grabs and kisses the spouse (with the mask still in place). When the passive asks, "What are you doing?" the psycho spouse does not reply, but simply continues embracing and kissing the passive (thus giving the passive a dose of his or her own medicine). Again, this may provoke an angry response, discussion, or wild love-making.

Caution: "Psycho Surprise" should not be played by older couples, or by anyone with a history of heart trouble, strokes, or the like. And it should only be used in connection with ongoing couples therapy, under the supervision of a psychotherapist. In some cases, for caution's sake, the "psycho" spouse may warn the passive spouse that a shower surprise might be in the offing at some point in the future—just to play it safe.

4

Games for Depressed Couples

Nearly every couple is depressed some of the time; hence the following games might be useful to all couples. Those couples who suffer from a long-term depression will benefit most, although they will also encounter the greatest amount of difficulty in getting out of their defensive postures.

What is the defensive posture of somebody who is depressed? It can be simply stated as "Why bother?" When you are depressed, you do not feel like doing anything—and that includes having sex. Getting out of bed in the morning can be a chore. Eating is more of a chore. Work is a terrible chore. Existence seems pointless and life empty of meaning.

It may be that some depression is due to an organic deficiency (as some researchers claim). However, it can easily be observed that much depression is the result of environmental conditions. If we lose a loved one, get fired from a job, or find ourselves evicted from our apartment, we become depressed. In infancy, the loss of a parent, a sibling, or even a treasured pet or doll can also cause depression. So can an array of other circumstances. If such childhood depressions are not successfully soothed by parents, the depression may remain as a character formation, so that as adults the slightest adversity can release the repressed infantile depression in the individual.

Depressed people negate life. They have felt negated as a child (or in some more-recent period), and now negate themselves and others. This serves a defensive function of protecting them from negation—for if they negate themselves as well as others before they negate them, they will be spared the pain of further negation. A secondary gain of depression is that it wins sympathy from others. However, generally depressed people reject sympathy. They cry out for it, only to reject it when it is offered. This repeating pattern expresses extreme ambivalence. It is also a reenactment of what happened during some earlier traumatic period.

One of my patients continually negated me. He would come late, pay late, and fall asleep during sessions. If I said anything to him, he would denigrate it, calling it "stupid." He denigrated my office and me alike, saying my furniture was cheap and my clothes tacky. If I tried to talk to him he would say he did not feel like talking, and if I tried to help him he would say nobody could help him. He also negated everybody else in his life. As soon as he had sex with a woman, he would feel disgusted by her and never want to see her again. Anybody who chose him for a friend was eventually seen as worthless.

As an infant, this patient was rejected by his mother, who seemed to have been a depressed personality herself. He was born with big ears and crossed eyes, and his mother used both unfortunate inheritances as excuses to be disappointed in him. Eventually his father and sisters also scorned him, cementing his development of low self-esteem and lifelong depression.

When two people like my patient get together, you generally have two people who have suffered from some kind of traumatic loss or some kind of emotional abuse. Naturally, they then displace their depression onto their primary relationship. Having been made to feel unworthy, they make their spouse feel unworthy; having not been soothed adequately, they have no sympathy for their spouse; having been emotionally abused, they emotionally abuse their spouse; having been deprived of attention, they are themselves depriving.

The games in this chapter are designed to deal with the negation that such people continually act out.

Game 1: The Fairy Godmother

Players: Depressed husband and fairy godmother (non-depressed wife).

Activist: Nondepressed wife.

Setting: Home.

Aim: Draw depressed spouse out of his depression by appealing to a rescue fantasy.

Game Plan: This game may be played with or without the depressed husband's knowledge, according to the nondepressed wife's discretion.

One evening, while the depressed husband is moping around the house, the wife rings the doorbell. She is wearing a revealing fairy godmother costume of silk or lace, in a color that will appeal to the husband, with a crown of some sort and a flowing cape. She holds a golden wand in her hand. If she has small breasts, she may enhance her costume with falsies. The object is to be slightly outrageous, but in a charming way. When the husband opens the door, she says, blinking her eyes theatrically, "Hi, I'm your fairy godmother!" She prances into the room, whirls around, and glides toward him. "Yes, yes— I'm here, I'm here!" she breathes in a soft, sexy voice. "You called, and I'm here!"

She guides him to a sofa or bed, sits beside him, and tousles his hair. "There now. Just lie back and let your godmommy take care of everything." She lays him back. "Yes, I'm here. Don't worry about a thing. I'll save you. Your fairy godmommy will save you. Do you like your fairy godmommy?" She sits up and wiggles sensually. "Sure you do," she coos sexily. "All little boys like their fairy godmommy! Now, you just relax. That's a good boy. Relax."

The fairy godmother may now try one of several approaches, depending on the nature of the husband's depression and whether he is in on the game. She may try the understanding approach, holding his hand and giving a prepared speech that utilizes things she knows about him. "You need someone to talk to. Somebody who'll really listen, who'll really listen perhaps for the first time in your life. Somebody who'll really be there for you, and hear all your complaints—no matter how stupid or asinine. All your life you've been looking for that certain person, that certain woman who would recognize your specialness and soothe all of life's unfairness. When you were a boy you used to fantasize about a fairy godmother like the one who rescued Cinderella, who would discover you and take

you away with her to a magic palace. When you were an ado-
lescent you used to fantasize that your English lit teacher
would find you and soothe you and make love to you. When
you were in college it was your American history teacher. I'm
all of these in one, and I'm here at last, ready to hear you and
care for you and make love to you as you've never been made
love to before. But first, tell me everything that's troubling
you."

If he demurs and does not launch into a recitation of his
woes from childhood on, then she might try the playful
approach:

"So, here I am!" She tickles him in a place she knows he's
vulnerable. "What do you think? Do you want to play with
me? Listen, I have a riddle for you. What's a zebra?" She sticks
out her breasts. "Give up? A zebra is twenty-five times bigger
than an A-bra!"

If he doesn't start laughing and cheering up (he may, for
example, become grouchier), the sexy approach may work:

"Hi there, handsome. What can I do for you?" She slides
her hand up his leg. "I'm here to relieve you of all your world-
ly and sexual tension, and I'm ready to fulfill your innermost
fantasies. Your wish is my compulsion!"

If the husband is in on this game, he will now find some
way to play along. If the game is a surprise, the fairy godmother must keep trying until she finds the key to unlocking
his resistance. That key usually turns on an understanding of
his particular rescue fantasy (all of us have one). Once it has
been found, the husband can be lured out of his depressive
posture and into an enjoyable sexual experience—which may
also lead to an unburdening of himself in a way he has not
experienced before. And this could in turn lead to increased
intimacy.

As with other games, this one must be played with conviction and zest. If there is any hesitancy, self-consciousness,
or inhibition, that will sabotage the proceedings. Therefore, the
active partner must be ready to truly throw herself into her
role and enjoy it. This will have a therapeutic benefit on her
too, channeling into a constructive groove her resentment
about her husband's depression.

Game 2: The Fairy Godfather

Players: Depressed wife and fairy godfather (nondepressed husband).

Activist: Nondepressed husband.

Setting: Home.

Aim: Draw wife out of her depression by appealing to her rescue fantasy.

Game Plan: The wife is moping around when the doorbell rings and she opens the door (or, alternately, the wife is lying in bed and the husband bursts into the room). He wears a costume that befits her fantasy—Superman, Robin Hood, a prince, a fairy with wings.

"Hi! It's me—your fairy godfather! My card!" He hands her a home-made card, then whirls around the room, his cape or wings flowing. Depending on the nature of his wife's depression and personality, he may dance around the room for a time, waving the magic wand, or stride toward her in a princely fashion.

"What are you doing?" his wife may ask in a sarcastic tone. "Stop being stupid."

If she is in on the game, she will play along of her own accord. If the game is a surprise, she may continue to try to negate it. (All such negation should be firmly countered.)

"Yes, dear, it's very stupid—but here I've come to rescue you. Would you please just let me rescue you? Please, just this once? I promise I won't hurt you. I know it's stupid and silly, but please let me be your fairy godfather and see what happens. What's up? Nothing's up. I'm just trying something— okay? Just lie back. That's right. Fairy goddaddy's going to make it all better. Sure he is. Lie back and relax."

As in the previous game, depending on the nature of her depression and on whether or not she has been prepped for his routine, he will take one of three approaches: (1) The understanding approach: "Now, dear, talk to me. I'm here to listen to you as you've never been listened to before!" (2) The playful approach: "Ha, ha, ha—does that tickle? Well, how about that? Now, here's a riddle for you: Why did the chicken stop halfway across the road? Give up? Because she wanted to lay it on the line. Get it? Ha, ha, ha!" He tickles her again. (3) The sexual approach: "How do you like your fairy goddaddy, my dear?" He wiggles and winks at her, touching his private parts. "Do you like his magic wand? Would you like his magic wand inside of you?" If all goes well, she will accept this offer.

The husband should use his imagination, embellishing the game with his own customized jokes, maneuvers, and the like. The game may be played more than one time. Just as a play gets better as it is rehearsed more, this kind of game improves with repeated performance, as inhibitions resolve. The first go-round may seem silly, and due to this may well be subsequently performed stiffly. But if both partners get into it and imbue it with their own particular rescue fantasies, the game will take wing and lead to better sex and better communication.

Game 3: Massage Poker

Players: Depressed husband and depressed wife.
Activists: Both husband and wife, by turns.
Setting: Home.
Aim: To relax body and mind, assuage depression, and release sexual and creative energy.
Game Plan: Husband and wife begin by playing poker, which quickly becomes strip poker: Each time one of them loses, he or she has to take off an article of clothing. The game ends when one of them is completely naked. The winner then has a choice of either giving or receiving a thorough massage.

The person who is to *give* the massage takes charge from that point on. Let's say it is the man, and that he then says, "You are about to have the most extraordinary massage you've ever had in your life. It will be an erotic, soulful massage, and you are to enjoy it to the fullest, and to try not to think about anything else but the massage. If you do think of something else, then always return to the massage and think about *it* again. If you want to think about how glum your life is, then go ahead and think about how glum life is, but then return to the massage again and think about that. If you want to think about how meaningless your existence is, then think about how meaningless your existence is, but then return to the massage again and think about that. Always return to the massage. Do you understand?" The giver of the massage should repeat this message until the receiver obviously does understand.

After undressing his wife, the husband takes her by hand into the bathroom and helps her into a hot bath and bathes her (slowly and lovingly) with a sponge. He then helps her out and dries her (again, slowly and lovingly), then leads her to the bed. She lies face down, and he begins the massage.

He lifts one leg and holds the foot up to his mouth and sucks on each toe, one at a time. He asks, "Is that all right?" If the answer is yes, he continues. (Should she say no, he will skip the rest of this part.) If he continues, he licks the arch of the foot. Then he lightly bites the heel. Then he gently lays that leg down, lifts the other, and sucks on each toe of that foot. Then he licks that arch, lightly bites the heel, and gently lays the leg down.

Next he lifts the right arm and holds the hand up to his mouth. He sucks the fingers and thumb of the right hand. He asks, "Is that all right?" (If the answer is yes, he continues; if no, he skips this part.) If he continues, he licks the palm of the hand and bites the heel. Then he gently lays that arm down and lifts the left hand, sucking each of the fingers and thumb. He then licks the palm, bites the heel and gently lays the hand down.

Now he sucks on the right earlobe, runs his tongue around the edge of the ear, and gently bites the top of the rim. Then he asks, "Is that all right?" If it is, he blows lightly into the ear. He then sucks on the left earlobe, runs his tongue around it, and gently bites the top of the rim.

Then he gently bites the calf of her left leg, and asks if *that* is all right. If so, he licks the area in back of the left knee. Then he bites the calf of her right leg. Then he licks the dimple in back of the right knee. Then he bites the back of her left thigh, then the back of her right thigh. Then he bites her left buttock. Then he bites her right buttock. Then he licks her spine, from her waist all the way up to her neck to where her hair starts.

If by any chance she does not like any of this, then he should persuade her to try it anyway, even if it feels strange or offensive—for only by trying it will she reap the benefits of the massage. It is her depression, he should tell her, that is offended by the massage, for it does not want to receive pleasure, does not feel worthy of it, needs always to negate everything. To overcome this depression, she should go with the massage—even it if does not at first seem enjoyable.

The next step in the massage (if he has persuaded her to go on) is to take a few ice cubes, wrap them in a face cloth or clean handkerchief, and slowly slide that along her spine to the nape of her neck. Then he slowly, lovingly slides the cold compress all around the shoulders, the back, the arms, the underarms, and the palms of the hands; between the fingers; all over the legs and the bottom of the feet; between the toes; up the inner leg and inner thigh on one side and down the inner thigh and inner leg on the other; then up the crevass between her buttocks, and circling around each cheek; then up the spine again; and finally around the sides of the head, across both the left and right temples.

Next he takes a hot "washrag" (dipped from time to time into a bowl of hot water) and slides it over her body, following the very same route as was used with the ice-filled cloth.

Next he takes a new feather duster and glides it along her body, following the same route as in the previous examples.

Next he begins slowly to kiss her body. He kisses in little pecks, again following the same route.

Next he takes something soft, such as a velvet or silk comforter, and covers her with it, slowly and lovingly rubbing the comforter all over her body.

Next he lies on top of her and envelops her with his naked body. He gently slides around, holding her hands in his as he does so, and kissing the back of her neck.

Then he asks her to roll over.

He licks in circles around each of her breasts and asks, "Is that all right?" If so, he continues. He licks each of the nipples, and bites each one gently. Then he runs his tongue down to her belly button and licks around it—then slides down to her vagina. He licks around her vagina and asks, "Is that all right?" If so, he continues. He licks her inner thighs, then bites each knee, and then each elbow.

He then takes the ice bag and runs it slowly about her body. He circles her breasts, then grazes each nipple. He runs the bag up her neck and around her right temple, over her forehead, down her left temple, very lightly across her lips, down her chin, around the rims of both ears, around her breasts again, down each inner arm, across each palm, then down to the belly button, around it, then down to the pubic hair, around it, down to the vagina, around it, then gently across it, then down the inner thighs, inner calves, and over the top of each foot.

Next he takes the hot washrag and follows the same route.

Next he takes the feather duster and follows the same route.

Next he kisses her, following the same route.

Next he covers her with a comforter and gently rubs her all over.

Next he lies on top of her, holding her hands, and rubs his body against her. Then he kisses her softly on the lips.

Then he asks, "Is it all right?"

If so, he asks, "Would you like an inner massage now, madam?"

If the answer is no, he stops the massage right there. If the answer is yes, he proceeds to fondle her, making sure she is wet, and then enters her. Looking at her and keeping her hands in his, he begins to have intercourse. He says, "It's very important to have an inner massage every so often. Try to think only of the inner massage now—but if you do think of something else, then gently return to the inner massage and think of that again. If you start thinking of how glum life is, then go ahead and think about how glum life is but then return to the inner massage and think of that. If you start thinking about how meaningless your existence is, then go ahead and think about the meaningless of existence and then return to the inner massage and think of that. Are you thinking about the inner massage now? Is it all right? Are you feeling better and better? It's all right if you want to come. I will protect you while you come."

The husband continues to give the inner massage until the wife reaches orgasm, or until she says she has had enough (she may not be able to reach orgasm this time around).

After the massage, the husband asks the wife to speak about her thoughts and feelings during the massage, encouraging her to be honest about any negativity that might have crept in.

This game can be repeated many times, with as many variations as the imaginations and inhibitions of the lovers allow, with ever-increasing results. However, because depressed couples are likely to be quite sexually blocked, results will be gradual. If done with caring, though, the game will be beneficial to both giver and receiver, the former getting in touch with his or her creative energy and learning to truly care about and give to his or her spouse (toward whom he or she may have been resentful or emotionally distancing), and the latter getting in touch with his or her body tension and emotional blocks (which kept him or her captive to depression). (Note: When the female is the massage-giver, she follows the same procedure.)

Game 4: Hopeless Sex

Players: Depressed spouse and nondepressed spouse.
Activists: Both.
Setting: Home.
Aim: To join and mirror the depressed spouse's defensive posture in order to facilitate him or her in developing more insight.
Game Plan: The depressed spouse has been rejecting all sexual overtures for some time. Let us say that in this case the depressed spouse is the husband. The wife now tries a different approach. She asks the husband to play this game with her.

They sit facing one another on a bed or rug. She fondles him to get him aroused. He fondles her too, although indifferently.

"It's hopeless," she says.
"Everything's hopeless," he says.
"Sex is hopeless," she says.
"Why bother?" he says.
"I'm too depressed," she says.
"I'm too depressed, too," he says.
"It's all so hopeless," she says.
"Why bother?" he says.

"What's the use?" she says.

"I won't get an erection," he says.

"I won't get wet," she says.

"I'll come too quickly," he says.

"I'll end up frustrated," she says.

"Why bother?" he says.

"Yes, it's hopeless," she says.

They go on, repeating variations of these phrases until they are ready to have sex. Then she climbs on top of him and begins to have sex with him. During the sexual experience, they are to look glumly into each other's eyes and say the negative litany again, exaggeratedly.

"Everything's hopeless."

"Yes, everything."

"Sex is hopeless."

"Absolutely hopeless."

"Nothing matters."

"Why bother?"

As they continue, they may become more personal.

"Life is hopeless, but I'm having sex with you anyway."

"It's all hopeless, but I'm kissing you."

"You're hopeless, but you have a nice tongue."

"You're hopeless, but I might as well use you." As they become excited, they continue the negativity.

"I'm getting excited, but it's useless."

"I'm getting a little excited too, but it doesn't mean anything."

"It's hopeless."

"Everything's hopeless."

"Perhaps I'll have a hopeless orgasm."

"Yes, a very hopeless orgasm."

The success of this game depends on getting both members out of their customary mode of relating, in which the depressed spouse negates both himself and his mate, and the nondepressed spouse continually tries either to soothe him or expresses resentment toward him. In this game, both accept and go along with the depression and the underlying feelings of hopelessness. Further, mirroring the depressed spouse's hopelessness gives him a glimpse of how he is acting. If they can both accept the depression and allow themselves to have hopeless sex, then they can move on and actually have hopeless sex. Then, ironically, they will find that the sex becomes less hopeless. It may also lead to getting more in touch with the hopelessness and letting go of it.

Game 5: Sexual Battle

Players: Husband and wife.
Activists: Both.
Setting: Bed.
Aim: Activate unconscious feelings of competition or resentment and direct them into a constructive channel.
Game Plan: Sometimes depression is involved with both feelings of oppression and an inhibition of self-assertion and competition. We learn as children that it is not all right to assert ourselves toward or compete with our brothers, sisters, or parents, for we will then meet with disapproval, condemnation, and the like. Hence the habit of allowing ourselves to be oppressed gets started.

In this game, a couple is asked to turn sex into a competitive sport (which at first seems opposed to everything that sex represents—i.e., the saying that was prevalent during the 1960s, "Make love, not war"). However, in this instance the competition, combined with eroticism, is being utilized to get at a particular kind of block related to a deep-seated fear of assertion.

The two begin as they normally would when making love. When they are aroused, they sit or lie opposite one another and begin to bring each other to orgasm using either hands or mouth—whichever they deem best. (Hand sex may be best for couples who have problems accepting oral sex or with achieving orgasm through intercourse.) In either case, as soon as the sex—of whatever kind—begins, the race is on. The partners set about trying to make their partner come first, and so each, of course, tries to resist letting go. The first person to achieve orgasm loses—and thus, of course, the person who causes the other to come wins. To spice up the game, the winner may get a prize—i.e., he or she will be the other's slave.

This game will provoke feelings that have lain dormant. Some people, when asked to make sex competitive, will scoff.

Others will suddenly have orgasms with a vengeance, whereas they previously had difficulty in obtaining them. Still others will take great pride in getting their partner (their opponent) to come first. In addition, the game puts each person into a conflict: to win the game the participants must try to make their mate come, yet on another level, the one who has the orgasm wins. Hence, either way they are both winning and losing. The person who comes first wins by losing; the other loses by winning. This conflict is not present in only this game, but also underlies the sexual block itself, and is unconsciously present whenever the participants have sex. All this game does is bring the conflict to the surface.

This game can be played again and again. It can be a fun game, with the participants making faces and using various other means (kissing, fondling, talking dirty) to hasten their partner's "defeat." However, there are no real winners or losers here—only two people who are putting authenticity and fun back into their sex and rekindling their desire.

"Sexual Battle," as with all other games described so far, will be successful to the degree that a couple can play it with sincerity. This goes without saying about any endeavor—you cannot cook a good meal without sincerity. However, when one is dealing with sex, people's resistances multiply, and there is a natural tendency to be critical and perhaps derisive. Depressed people may want to negate this game (and all other games in this book), finding them silly or too complicated or too simple. I have found that I have had to give my own clients lengthy pep talks to properly orient them to the games. Because these are not miracle-workers, but only devices to help couples get closer, they are only as effective as the couples who play them allow them to be.

5

Games for Hysterical Couples

Actually, there are no hysterical couples. There are hysterical people, and such people tend to dominate their partners through their hysteria. Hence, they make their relationships hysterical.

Hysterics are generally women, although in some cases they are men who have a strong identification with a hysterical parent. Such people tend to be emotional and unreasonable—particularly about sex. Hysterics think that sex is dirty, disgusting, exploitative, and an invasion, even when it involves their spouse, the one who is presumably the love of their life. They will defend against these feelings in several ways: by controlling the sexual act to such an extent that they psychologically castrate their mate; by rejecting their spouses's advances entirely; or by keeping the negative feelings repressed and developing a *reaction formation*—that is, becoming obsessed with sex.

I once treated a hysteric who could have sex only if she had fantasies of being ravished—sometimes by a whole regiment. This peculiar defensive posture was developed in childhood to protect her against the onslaught of sexual rejection or shaming by her parents, sexual competition with her mother, and inappropriate sexual advances from her father, uncles, or brothers. Now it simply keeps her separated from, and in opposition to, her husband.

Generally, the hysterical couple is composed of a hysterical female and a passive male. Such a husband usually is a man who had a hysterical mother and is now attracted to a hysterical woman in order to gain "mother's" approval. He thinks that if he gives in to the hysteric's unreasonable demands, she will approve of him and he will get what he wants—her love. By submitting constantly to her unreasonable demands, he will, of course, get lots and lots of approval but no sex (or much more sex than he can handle) and little real love or respect. In fact, one of the only ways a female hysteric can enjoy sex without guilt is if she is "swept off her feet," the husband taking full responsibility for the encounter. Ironically, however, the kind of man she chooses as a husband is someone who cannot do that, since his need for her approval precludes his doing anything of which she would consciously disapprove.

On occasion, hysterics marry active spouses, hoping that such men will sweep them off their feet and take them away from it all. But the actives are more than likely to be of the narcissistic variety, interested in satisfying their own needs, not their spouse's. For this reason, many such relationships swing from fantasy to disillusionment.

The games in this section have been designed as a five-part antidote to hysteria.

Game 1: The Kissing Bandit

Players: Hysterical wife and passive husband.
Activist: Husband, with or without wife's knowledge.
Setting: Home or vacation retreat.
Aim: Help husband overcome need for wife's approval by learning to assert himself; stimulate wife's sexuality by appealing to her unconscious fantasy.
Game Plan: The couple can play this game together, or the husband can spring it on his wife if he is sure she will respond appropriately. They can play it at home or during a vacation— say on a Caribbean island. (Being at a very different, perhaps even exotic, locale adds a bit more spice to the game.)

The husband dresses up in some kind of bandit's costume, such as a cowboy outfit with a mask à la the Lone Ranger, or a pirate's gear—again with a mask of some sort. Then, while his wife is lying in bed or sitting in the living room reading or watching TV, the husband breaks into the room. He stands before her silently, arms folded.

"Who are you?" she may ask—or, "What are you doing?"

"Never mind," he will answer in a brave new tone of authority. "Put down that magazine [turn off the television] and come with me."

"Where?"

"Never mind where. Just do as I say."

"No. I'm reading my favorite book [watching TV]."

"I said, put it down [shut it off]."

"Stop being so silly. Where did you get that costume?"

"I'm not going to ask you again."

"Stop talking like that. What are you trying to prove?"

"I'm not trying to prove anything. I have nothing to prove. I am who I am. I am the Kissing Bandit, and I may just steal some kisses from you."

"Oh, yes?"

"Yes!"

The conversation goes on like this for a while, perhaps. Then the husband walks up to the wife, tosses the book aside [turns off the TV], and takes her into his arms. If he is physically able, he lifts her up and cradles her in his arms. If not, he pulls her to a standing position and leads her by the arm.

"Where are you taking me?"

"Where I should have taken you long ago!"

"Put me down [let go of my arm] this instant!"

"No."

"I said put me down."

"No!"

If the husband and wife are playing this game together (cooperatively), they may make up their own lines at this point. If the husband is the activist, he should proceed with caution: If his wife is smiling with surprise or delight at this sudden and somewhat fanciful act of assertion, he should continue. If, however, she is adamant about not wanting to be swept away, he should not press on. Instead, he should try, in as authoritative a manner as possible, to persuade her to go with him. "Why don't you just do as I say and see what happens? You might just be pleasurably surprised. Come with me. Is our life so great that we couldn't use a little mystery, a little surprise? Just this once let me take command. I promise you I won't do anything to hurt you, and you may just find yourself enjoying it." If all else fails, he should abandon the attempt to be the sole activist, reveal the game to her, and encourage their mutual participation in it from this point on.

The husband then carries (or leads) the wife into another place (another room, a basement, a friend's apartment) where he has set up some kind of "cove" or "hideout." Perhaps there's a mattress on the floor, or the bed has a different bedspread on it (such as black satin), and there's a treasure trunk or box near the bed (containing a .new bauble for his captured princess), and a bottle of champagne and two glasses. Some exciting "bandit music" pulsates from the stereo—Ravel's *Bolero* comes to mind—and an exciting aroma comes from incense. He lays her down on the bed, pours some champagne, and offers a toast.

"To my new princess!"

"Champagne? I don't believe you."

"Drink it."

"I don't want it. Why don't you take off that silly mask?"

"I said drink it."

"What are you going to do with me?"

"Something I should have done long ago."

"Are you going to ravish me?"

"Like you've never been ravished in your life."

"You masked brute. Unhand me!"

"You're free to go any time you want."

"I am?"

"Yes. But if you do stay, be forewarned that I shall spare nothing in my attempt to please you sexually."

"Nothing?"

"Nothing."

He envelops her in his strong arms.

"You're holding me so tight."

"That is correct."

He kisses her hard. At first she playfights to get away, then coyly surrenders to his new force and verve, and kisses him back. If the husband and wife are playing this together, they will have no trouble making up their own variations. If the husband is still the sole activist, he needs to proceed with caution: By now, even if the wife was initially surprised, she will probably be playing happily along with him. However, if she is frightened or angry or feels threatened, he should not press on. Instead, he ought to use the occasion to talk about both their feelings. This in itself can be fruitful if he finds out what it is about this game that has upset her. Such questioning is certain to open up new avenues of communication about the sexual realm.

Otherwise, the game continues. The husband tosses the empty champagne glasses aside (or breaks them in a fireplace, if feasible) and says, "Oh—here's a little something for you. Put this on."

She opens the treasure box and finds a necklace or bracelet.

"For me? Really?"

"That's right."

She tries it on and primps before a mirror. "So you think you can buy my love?"

"A bandit doesn't have to buy anything. He steals it."

He takes her into his arms and steals a kiss.

"Please," she murmurs. "Be gentle."

He undresses her and himself, then makes love to her in a more authoritarian way than ever before. He can do this by using a different variation of foreplay (perhaps doing something she has requested but he has heretofore felt squeamish about), insisting on a different position, or just being a little more forceful. Afterwards, the couple will find that this game has either invigorated their sex life or has brought up feelings that stand in the way of a better sex life. Even in cases where the game at first seems corny and either the wife or husband abandons it, good results can eventuate. The game will put the husband in touch with both his fear of self-assertion and his need for his wife's approval. It will put the wife in touch with her fear and loathing of sex and her deep-seated need to be swept away so that she doesn't have to feel guilty about doing something "dirty."

Game 2: Nude Indoor Volleyball

Players: Husband and wife.

Activists: Both.

Setting: Large room, such as living room, den, basement, or garage.

Aim: Rekindle a playful attitude toward sex.

Game Plan: All of us, as children, once had a playful attitude toward sex. Indeed, in psychoanalysis we say that sex is regression in the service of the ego. When sex is going well, we become not only childlike but even infantile, expressing all the pent-up needs from the earliest past to the present. This game serves to directly facilitate this playful attitude toward sex and foster a regression in the service of the ego.

Tie a volleyball net, rope, or string across the middle of a large room. (Whatever it is, call it the net.) Mark boundaries of the minicourt with a tape on the floor, or use a 9' x 12' rug as the court surface. Blow up a balloon to the size of a volleyball; the balloon is the volleyball.

As in regular volleyball, the server stands behind the line to serve, and scores points only on his or her serve. After the serve, each player has three hits to get the ball over the net. (One hit can be used to block the ball, one to set it up, one to spike it, etc.) Play stops when a player either fails to get the ball back over the net or knocks it out of bounds. The first player to reach 15 points (one point per serve) wins.

However, this brand of volleyball has two distinct twists. First, it is played in the nude. Second, the players are allowed to try to distract one another's play by reaching under the net to fondle the other's genitals while the opponent is trying to make a play. If you can arouse your opponent to such an extent that he or she misses a ball, great. If both become so aroused that they forget about playing entirely and fall to the floor in heated passion (perhaps popping the balloon on the way to the floor), that's even better.

This game circumvents in several ways the hysteric's aversion to sex. First, it takes the pressure off of her to have sex with her husband: Instead of trying to persuade her to have sex, he invites her to play a game. Second, it bypasses her resistance to sex by veering her toward Eros indirectly, through play, as well as by appealing to her competitiveness toward men (generally a component of hysteria). Third, as mentioned earlier, it fosters a spirit of play (or, as we used to say, of "good, clean fun"), taking the experience out of the realm of sexual exploitation or disgustingness—which, on a deeper level, is where the hysteric places it.

For the husband it is a transforming experience, lifting him out of his own defensive posture—which might range from whining about lack of sex, to passive but begrudging resignation to little or unresponsive sex, to seeking extramarital situations and satisfactions. By dropping his defensive posture and allowing his own playful self to come out, he learns a more successful mode of relating.

Nude indoor volleyball can be played for fun and enlightenment by most of the couples described in this book.

Game 3: Prostitute

Players: Prostitute and john.
Activist: Wife, without husband's foreknowledge, or both.
Setting: Home or hotel.
Aim: Arouse wife's sexual passion by appealing to her prostitute fantasies, while prodding husband out of his oral ("Take care of me, please!") passivity.
Game Plan: If the wife is up to it, she may want to spring this game on her unsuspecting husband some night at home or away during a vacation. Or, they may both participate in setting up and playing the game. If the wife activates the game on her own, it is more likely not only to plug in to her whore fantasies, but also to appeal to the angry, proud aspect of her character.

One night while the husband is either home or in a rental room alone, the wife rings the doorbell several times, insistently. He opens it to find a quite different wife than he has ever seen before. She is dressed for the part, with hair all askew, oodles of red lipstick, eyeliner, rouge, a low-cut neckline, a miniskirt, net stockings, and high-heeled shoes. She smiles seductively and slithers saucily across the room.

"Do you want me?" she asks.

"Do I *want* you?" the husband may ask in surprise. "Yes. Yes, I do want you."

"How much are you willing to pay?"

"Pay?"

"Yes. I'm for hire."

"You're for hire? What are you doing?"

"What does it look like I'm doing? I'm a hooker. I'm your hooker for a night."

"Are you serious?"

"I'm dead serious. Care to sample my goods?" She flashes a breast.

"You'll do anything I ask?"

"Yes. Within reason, that is. And for a price."

"What's the price?"

"That depends on what you want. Straight sex is $200. If you want something extra, that'll cost you more."

"Extras? Like what?"

"Well, like if you want a striptease first. Or if you want me to do something unusual—play out some fantasy."

"Hmmm."

"I'll tell you what. I'll throw in the striptease for free."

"That's nice of you."

"So, do we have a deal?"

"I guess so. I'm not sure what you're up to, but I think I'm going to like it."

"That'll be $200."

"I have to pay you now?"

"I prefer that clients pay up front. That way there's no problem later. Not that I don't trust you."

"I understand."

The shocked but delighted husband will probably pay the $200 without a hitch. If he isn't in on the game, he may even be flabbergasted and mutter something like "What about your headache?" or "What's come over you?" Or he may bicker about paying for his wife's sexual favors. If he bickers too much, the wife may counter by handing him some sort of made-up business card and saying, "All right—forget it. But if you change your mind and wish my services, use this." She may then walk toward the door. Either the husband will stop her, or he may take her up on the offer at a later date, after he has thought it over.

Once the husband has paid (it is important to have him pay first, otherwise some of the effectiveness of the game will be lost), the wife proceeds to do a striptease, using her own imagination. Then she sits on the husband's lap.

"So, what would you like now, my dear?"

"You'll do anything?"

"Anything you like."

They proceed according to the husband's fantasies. Having been paid for her services, the wife will give herself to the experience with a new zest, and the husband will in turn respond similarly.

If the game goes well, it should not only enliven their relationship, but also serve to play out each of their deeper repressed fantasies. By giving vent to these fantasies, they break through the block that has kept their sexual relationship at a stalemate. These fantasies are attached to repressed feelings of anger having to do with early-childhood sexual traumas, and may therefore include "rough" talk or action. After the sexual part of the game is finished, the couple should talk about how it felt to play the game, and what that says about the way they had related up to that point. They may repeat the game as often as they like—or even try a reversal of the game in which the husband plays the prostitute and stripper.

Game 4: Reverse Headache

Players: Husband and hysterical wife.
Activist: Husband.
Setting: Home.
Aim: Use of paradoxical mirroring of wife's headache maneuver by husband in order to provoke an authentic confrontation.
Game Plan: Like the passive, the hysterical woman gets headaches—often migraines—and uses them as an excuse to refuse sex. However, the hysteric's headaches are more severe and are sometimes accompanied by fits. "I said I have a headache, and I mean I have a headache! Don't you have any

consideration or respect for me at all?" The angrier variety of hysteric, therefore, will not respond to the game called "Headache" (see chapter 3) nor to any pleas, demands, or discussions. That type needs a more forceful brand of emotional communication.

In this present game the husband imitates the wife's behavior in a way somewhat like what children do to one another. He does not ask for sex from her anymore but rather waits until she asks something of him and then refuses—saying "I have a headache." It can be a small or large request by her which elicits this response. For example, they may be watching television and she may say, "Would you change it to 'Wheel of Fortune'?"

"No, sorry—I can't. That show gives me a headache."

"Well, it's my favorite show, and I want to watch it."

"Don't bother me right now. I have a headache."

"Would you please switch the channel."

"No."

"Then I'll go watch it in the bedroom."

"Suit yourself, but I'm not switching the channel, because that show will make my headache worse."

Of course, the wife will be suspicious and may tell him she is not fooled by him—that she knows he is imitating her, and it is not going to work. But the husband should persevere despite her protests, bemusement, taunts, and the like, never letting on to her that he is playing a game, and always insisting that he really does have a headache. This is important, because if he lets up and admits he is just imitating her, the effect of the game will be lost. Indeed, this game (like every other paradoxical game in this book) relies on giving exacting performances that frustrate and wear down the "opponent."

The husband may have to continue this game for days—even weeks. Sometimes his "headaches" will occur with respect to small matters, such as emptying the trash, and sometimes with regard to bigger things.

"Why aren't you getting dressed?" the wife asks.

"I'm sorry. Going to the theater gives me a headache."

"You never got headaches when we went to the theater before."

"That's true, but I'm getting them now."

"You know, this game of yours is no longer amusing."

He places his hands on his temples and looks as if he is about to faint. "And I'm getting sick and tired of you not taking my headaches seriously. You think I'm faking, don't you, because I want to get out of going to the theater? I wish you could respect my feelings just once. Is that too much to ask? All you think about is your own needs—never mine!" (This little speech should be an imitation of one of her repetitious tirades brought on when he formerly approached her for sex.)

Eventually, if the husband can stay with it to the point where he becomes a nuisance, claiming to have a headache whenever the wife asks for anything, she will become upset. When she becomes so upset that she'll try anything to get him to stop (even going so far as to offer to have sex with him if he will relent about something he has been refusing to do), he has "won" the game. Having broken through her own resistance, she will now have a new respect for him and will experience sex with him quite differently.

"You bastard! You really think you're clever, don't you?" she may say as they are writhing on the bed.

"Perhaps a little."

"Admit it—you've just been mimicking me all along."

"Maybe."

"Maybe?"

"Shut up and kiss me."

"Anything you say, dear."

Only after he has sexually gorged himself—and her— should the husband confess that he has been playing a game with her, by way of retaliation. The wife may be ready now to confess that she had long been playing the same game in order to gain satisfaction through continually frustrating him. The husband will confess that he had become so angry at his wife that at times he wanted to strangle her, and that playing this game gave him an immense amount of substitute satisfaction. She will no longer be stuck in the hysterical/sadistic mode—and he no longer in the passive/masochistic mode. She may be resentful for a while at having her hysterical defense thus mirrored, but eventually the game will lead to a discussion (and perhaps to negotiation) by two increasingly respect-ful adults—and thus to a more mutually gratifying relation-ship.

This game may seem drastic or childish. One of my patients, to whom I suggested a variation of it, replied skepti-cally, "But that's playing games!" What he meant of course is that it was a kind of manipulation entailing "tit for tat" behav-ior. It seemed excessive and revengeful to him. "If I have to do something like that to get her to want me, then it wouldn't be real; it would be contrived." There is an element of truth to this complaint, I told him—but I make a distinction between playing a game simply to gain revenge and playing a game designed to end a stalemate and achieve closeness. The former I call a *subjective* form of acting-out, while the latter constitutes *objective* acting-out. The rationale of objective acting-out is to oppose subjective acting-out (the wife's headache maneuver) in the only way that will truly reach her—by acting out back to her. Sometimes you have to go to another person's level of game-playing before you can transcend it and reach more-authentic relating!

Another objection by this same patient was, "What if my wife isn't playing a game? What if she really gets headaches? Isn't it cruel to imitate her?" Generally when a symptom (headache) happens repeatedly and directly in response to the same situation (sex), it is psychosomatic. This is not to say that the headaches are not real. They are real—but they also are somatic representations of unconscious feelings. So to cure the headache entails resolving the feelings underneath. It is no crueler to mirror her headaches than it would be to apply an ice bag to her temple.

The other reason why my patient was skeptical was that he didn't want to give up the game he had been unconsciously playing until then—the passive game of "Oh, what I have to endure because of you." He would rather have continued that game and derive its secondary benefits of feeling morally supe-rior to his "rejecting" wife than take an action designed to end their ritualistic conundrum. All too often we would rather cling to an old way, even if it is not working, than venture to a new one.

Game 5: Talk Dirty to Me

Players: Husband and wife.
Activist: Husband, without wife's knowledge, or both.
Setting: Any bedroom.
Aim: Husband deliberately uses dirty language that wife consciously finds repugnant but unconsciously fantasizes about, thereby making conscious that which was formerly unconscious.

Game Plan: Some evening (or morning or afternoon, as the case may be) while the husband is making love to the wife, he suddenly looks at her and says,

"You slut."

"What?"

"You slut. You dirty little slut."

"Why are you saying that?"

"Because that's what you are—a dirty little slut."

"I am not."

"You are, and you know it. And don't pretend to be shocked by this language. You know you like it. A dirty little slut like you always likes dirty talk. And dirty sex, too—right?"

If the wife has a shocked but bemused smile, the husband continues to talk dirty to her, using language of his own choosing and letting the wife's response be his cue. He should experiment with different words and note which ones arouse her the most. For instance, he might say, "You really like my filthy cock inside your dirty little slit, don't you? Admit it." And if she nods emphatically and kisses him passionately, he should repeat that phrase several times, and then try others, developing a whole repertoire. The wife may then also begin trying some choice phrases.

If the wife becomes truly upset, however, the husband should discontinue the game and ask her to participate actively, explaining the benefits of this kind of game—how it is designed to appeal to her unconscious fantasies and be therapeutic. If she does not agree to play the game on the first occasion, the husband should keep trying.

When the game becomes more actively played by both, it takes off on its own. The dirty talk will have a liberating effect on both partners, since this forbidden language, as well as the ideation behind it, are being repressed and hence are blocking both true love and sex. The language and the sex alike become more abandoned with each playing of the game.

The final step is to discuss the significance of the dirty talk—what it means to both partners, how it "feels," and even where it comes from. This step is very important, for without it the game will simply be an enactment of the fantasies without resolving the block that creates the need for the fantasies. This reminds me of the male patient I wrote about in the book's Introduction, who was seduced by the young woman who wanted him to talk dirty to her while she sucked her thumb. While such enactment of one's sexual fantasies is gratifying, it is not therapeutic. Instead, it becomes a repeating pattern that feeds upon itself while never really achieving the ultimate satisfaction of a real connection.

Talking about the experience of using common language during sex—discussing one's need to talk dirty or to hear dirty talk—leads to insight about how one's natural feelings about sex got to be derailed and one's capacity for unhindered tenderness was blocked. It moves the experience out of the level of compulsive acting-out to a higher level of awareness, trust, and bond-building. When that happens, sex transcends the realm of ritual and becomes rich with a deeper meaning.

6

Games for Narcissistic Couples

Narcissistic couples cannot bond adequately because of problems of the self. They generally suffer from low self-esteem and compensate via self-absorption, pride, and grandiosity. They tend to become enraged if their pride is hurt and can get quite nasty to a spouse by whom they feel betrayed. Since they are often "stuck on themselves" (as was the mythological figure Narcissus, after whom this character type is named), they do not feel much empathy for mates, viewing them instead as objects to exploit rather than to love.

There are two basic kinds of narcissistic relationships. One, designated a *twinship transference* by Heinz Kohut, a psychoanalytic specialist in narcissistic disorders, is an alliance of two grandiose individuals who mutually support each other's proud and inflated self-perceptions while remaining blind to each other's delusions. They typically view themselves as a pair of pearls among swine—or (perhaps a bit more graciously) among synthetic pearls. The other kind of narcissistic relationship is called an *idealizing transference*. Here, a narcissist who feels inferior and needs to idealize somebody attaches himself or herself to a narcissist who feels superior and needs to be idealized. The former hopes that, by coupling with the latter, some of the "superiority" of the idealized object will rub off. The latter hopes to bolster his or her esteem by being the object of idealization.

We call these transferences because they contain—as indeed in one or another way all relationships do—an element of displacement; each member of a narcissistic coupling transfers onto the present relationship a quality of relating that existed in a primary relationship during early childhood. Thus, a boy who idealized his narcissistic mother will seek out a narcissistic wife, whom he will also idealize and in whose glow he will also bask. A girl who enjoyed a twinship relationship with her father, or perhaps with a brother, will seek out a similar relationship with a spouse.

The problem here is that if we are in a narcissistic relationship, whether twinship or idealizing, we do not truly bond with the other, but instead ally with an object who represents our ideal image of ourselves projected onto or mirrored by another person. Hence we never separate from the other (or from our parents) to form *our own* wholly independent self, and continually need to bolster our self-esteem in this way. Narcissists do not ask what they can do for others, but what others can do for them.

Furthermore, sexual relationships of narcissists have more to do with getting narcissistic needs met—merging sexually with a symbol of status, say—than with tenderness and love. The male narcissist, feeling doubt about his self-worth, is attracted to a partner who is most like himself (often in fact a mirror image of himself, as in homosexuality), one who in some way does that self proud. Thus a proud man will look for a proud woman. Conversely, a beautiful woman will look for a handsome or rich man. Their sexual relationship will be satisfactory only on a superficial level, and then only as long as their twinship or idealizing transference lasts. Of course, such transferences can become negative and even ignite rage and envy. An example of this in fairy tales is when the witch in *Snow White* says, "Mirror, mirror, on the wall, who is the fairest of them all?" and the mirror replies, "Snow White." The witch becomes enraged and tries to use witchcraft to destroy her rival.

The games for narcissistic couples are designed to counter and resolve their unhealthy narcissism.

Game 1: Cinderella

Players: Cinderella and the Prince.
Activists: The husband may activate the game on his own, or husband and wife may activate it together.

Setting: Home.

Aim: Husband appeals to wife's fantasy about meeting a prince who would discover that she is a princess.

Game Plan: One evening when he knows his wife "has no plans," the husband arranges to come home a little late. He dresses up in a prince costume and knocks on the wife's bedroom door. She may be just about to doze off.

"Excuse me. Are you Cinderella?"

"Cinderella?"

"Yes, Cinderella. The enchanting Cinderella. Because I'm Prince Charming, and I'm here to find out whether your foot fits into this glass slipper." He dangles a glass slipper before her surprised eyes. "Ah, yes—I've been searching far and wide for the young woman who danced so beautifully at the Queen's New Year's Ball. I've knocked on a thousand doors and tried this slipper on a thousand left feet. Whoever can fit into this slipper will be my bride and live with me forever in my castle in northern New Jersey. Would you care to try it on?"

"Oh, I get it. I'm supposed to be Cinderella."

"Aren't you?"

"Well, actually, now that you mention it, I *am* Cinderella!"

"I thought so."

"Where would you like my foot?"

"On this footstool would be fine."

"Certainly, Prince."

The prince sits on a chair or stool and takes Cinderella's foot into both hands, lovingly. He soaks a washcloth in a bowl of warm water, rinses it, and proceeds to tenderly wash around Cinderella's left ankle, foot, and toes (and even between her toes). Then he kisses her toes, the bottom of her foot, the ankle, the calf, the inner knee, and the thigh—and then works his way down again. He takes his time, preparing the foot for the ultimate fitting.

"Is this part of the fitting process?" Cinderella asks.

"Oh, definitely."

"I think I like it."

"I thought you would."

Perhaps because it is based on such a familiar story, this game will most likely not encounter much resistance. It is fun to play and appeals to the grandiose fantasies of both wife and husband. The game will in effect take on a life of its own. Cinderella will sportingly try on the shoe, and the prince will exclaim, "It fits! It fits! Now—you will be my lawful-wedded wife!" They embrace. From there the story is up to the participants. The prince may present her with a princess gown, and she may dress up in it and then actually go through a wedding ceremony. There could even be a wedding ball, with candlelight dinner and midnight dancing and brandy punch drunk from her slipper. Finally he carries (or leads) her over a threshold and right into bed.

As they are making love, he says to her, in exactly these words:

"Yes, I am a prince and I'm making love to you."

And she says:

"Yes, I am a princess and I'm making love to you."

And he says:

"I must be great if I'm making love to a princess."

And she says:

"I must be great if I'm making love to a prince."

They should repeat these and similar phrases during the course of the lovemaking. It will serve to emphasize their narcissistic feelings about one another while pointing to the lack of more genuine feelings. Naturally, this part of the game is rather less fun and more confrontational. Couples may well want to resist saying these things, viewing them as spoiling the fun or interfering with the flow of the story. But it is important to work through any resistance and express themselves. They can do so by discussing feelings of resistance as they come up. After the lovemaking they should talk about what other thoughts or feelings or memories they had while playing the game—and particularly while saying their designated lines as they were making love. A number of surprising feelings and memories will emerge that had formerly lain in the unconscious.

The game can also be reversed, with the husband playing Cinderfella, and the wife playing a princess looking for her mate.

Game 2: I Love You Just the Way You Are

Players: Husband and wife.
Activists: Both.
Setting: Home or hotel.
Aim: To arouse unconscious, unspoken judgments.

Game Plan: This is a simple game, but its results can be profound. It is for the narcissistic couple who started off in a twinship or idealizing transference but have drifted apart. Sometimes, in such cases, either there is little sex or it is rushed and orgasm comes too quickly or not at all. Sometimes one or the other feels disillusioned—even contemptuous toward the mate. Sometimes a quiet (and sometimes a loud) rage is expressed. Sometimes envy or possessiveness gets in the way of effective relating: The wife envies the husband's job and mobility, and the husband is possessive of the wife.

During a weekend evening or other relaxed occasion, the husband and wife undress and entwine themselves face to face. A good position for this game is for the husband to lie across a bed with his back and head propped on pillows, and the wife to straddle him. After he enters her and they are one, they should gaze into one another's eyes. Eye contact is important. They should try to maintain eye contact throughout the game and notice how they feel about that contact and what makes them want to look away. Once they have assumed their coital position, the wife says to the husband:

"I love you just the way you are."

Then she says the next thing that comes into her mind.

Then the husband says:

"I love you just the way you are."

Then he says the next thing that comes into his mind.

They repeat this simple sentence and the follow-up thoughts as long as necessary—that is, as long as it takes to say everything that pops up from their unconscious (or formerly censored conscious).

What happens is that whenever they say, "I love you just the way you are," the next thought that will come into their minds is a negative judgment, such as, "Except I wish you'd lose weight," or "Except you come from a poor family and don't know how to be rich," or "Except I hate the gap in your teeth," or "Except I wish you were a little smarter/prettier/a better dresser." As these negative judgments are acknowledged, the impasse will be broken, replaced by the feelings they have been withholding from one another. Sometimes arguments ensue:

"Oh, so I'm not pretty enough for you."

"No, you're pretty enough—it's just that I have these perfectionistic standards."

"Well, you're no movie star yourself."

"That's true."

If these initial reactions can be worked through in a spirit of good faith, eventually the couple will develop a more honest relationship, instead of pretending they still admire one another when they do not, and end up being hypercritical or silently sulking or gloating. As their relationship becomes more true, so does the sex. So, while each partner takes turns saying, "I love you just the way you are," and then says what comes up next, their sexual experience will take some surprising twists and turns. It may be angry, sad, rough, incredulous, and much less exploitative. They will say the sentence, say their thoughts, and then find themselves engaging in rougher sex than usual. Or they will say the sentence, say their thoughts, and cling to one another almost desperately, like infants. The impulse to use the other as a narcissistic sexual extension will diminish. This exercise can be done again and again, with new thoughts coming up, each time accompanied by new kinds of sexual feelings.

To go a step further, couples might also try a variation in which they take turns saying, "I love myself just the way I am," and then say the next thought that comes into their mind. This will allow them to trace back to its source the judgment that was being projected onto the lover.

Game 3: Freak Love

Players: Husband and wife.

Activists: Both.

Setting: Home or hotel.

Aim: Put couples in touch with excessive pride by forcing them to have sex with a "freak."

Game Plan: This is the opposite of the children's game of

"Dress-Up," the object of which is to make yourself look as beautiful as possible. In this game the couple make themselves as ugly as possible.

They set a time and place for a rendezvous. It could be at home or at a hotel. (Renting a room may work best, since it adds another exotic element to the game.) Before meeting, the wife and husband each dress and make themselves up so that they will appear as ugly and freaky as possible. For example, they may wear a grotesque witch or monster mask, or dress up in rags, or paint their faces oddly, or wear a wig, or put yellow gunk on their teeth, or douse themselves with awful-smelling perfume. They can cover their bodies with "green slime" (stuff sold in toy stores), or wear a pillow under their clothes to make it appear they have a hump or a fat belly.

The kind of freak one chooses to be will express unconscious fears. For example, men who harbor feelings that they are monstrous letches will dress that part. Women who have buried fears that they are evil witches will don the familiar witch's hat and wield a broom. By dressing up like the freaks they unconsciously fear they are, that part of their personality, formerly repressed, will be allowed to surface.

Making love, either in a hotel room or at home, will be an enlightening experience. They will not only be confronted with actually being the freak they fear they are, but also with making love to the freak they fear the other is. It will lead to interesting sex—and an equally interesting change in how they feel about themselves and their partner. To confront the freak within us is to stop projecting that freak on others!

Note: To enhance the effect of this game, participants should not only dress the part but also act it with conviction.

Game 4: Master and Slave

Players: Husband and wife.
Activists: Both.
Setting: Home or hotel.
Aim: Use paradoxical approach to get at couple's inferiority/superiority feelings.
Game Plan: The couple may take turns being master and slave. If they have a twinship transference, either can go first.

If it is an idealizing transference, the idealizing partner should be the slave first, since that order is closest to their present relationship. The taking of turns at being the slave leads to interesting results.

It may begin after a dinner at a fine restaurant, or in the restaurant of the hotel where they are staying. Upon coming back home, the master sits down on an easy chair and looks at the slave sternly. (In this example I will have the man play master.)

"Come over here."

"Yes, sir!"

"Kneel down before me."

"Yes, sir!"

"Do you realize that you are nothing and I am everything? Do you realize that you are just a slave and I am your master? Do you?"

"Yes, sir!"

"Do you realize that I know everything and you know nothing? Do you?"

"Yes, sir!"

"Without me you would be nothing. Nothing! Do you hear?"

"I hear, sir!"

"Nothing. That's why you have to stay with me. That's why you have to be my slave. You're only something at all because you're with me, who's everything. Say, 'You are everything, Master, and I am nothing.'"

"You are everything, Master, and I am nothing."

"Now, do exactly as I say!"

"Yes, sir."

"Unzip my fly. Unzip it right now."

"Yes, sir."

"Take it out."

"Take out what, sir?"

"Don't act stupid. Take it out."

"Your cock, sir?"

"Don't say that word."

"Yes, sir."

"Fondle it."

"Yes, sir."

"That's right, keep fondling it."

"Yes, sir. Does it feel good, sir?"

"Don't ask questions. I didn't say you could ask me questions."

"Sorry, sir."

"Now, take off your clothes. Leave on your panties. Follow my instructions exactly."

"Yes, sir."

"That's right. Take them off."

"Yes, sir."

"But leave on your panties, because I don't want to see your dirty hole."

"Yes, sir."

"Now, stand before me. Turn around. Bend over."

"What are you going to do, sir?"

"I said don't ask questions."

"Yes, sir."

"Do you like my finger in your hole?"

"Yes, sir."

"You would, an inferior little nothing like you."

"Yes, sir."

"Now, get over on the bed."

"Yes, sir."

"Lie back."

"What are you going to do, sir?"

"Didn't you hear me when I told you not to ask questions?"

"Sorry, sir."

"Now, I'm going to fuck you, even though you're inferior to me."

"Thank you, sir."

"My cock is too good for your pussy, but I'm going to put it inside you anyway. Once my golden cock is inside you, you'll feel as though you're almost as good as I am."

"Thank you, sir."

"Do you like that?"

"Yes, sir."

"Of course you do. Why wouldn't an inferior being like it?"

The game can have as many variations as the temperaments and limitations of the couple allow. At first it may seem awkward, but if the players stay with it they will be rewarded. All narcissistic couples benefit from playing this game, because it enacts both their grandiosity and the feelings of low self-worth for which the grandiosity overcompensates. Couples in idealizing and idealized transferences will find this game more natural. When the idealizing partner plays the slave and the idealized partner the master, they will feel a bit embarrassed since it hits so close to the truth of their situation. However, when they reverse roles, both may have an even harder time playing the game. This is also true when a couple with a twinship transference plays it. Those prone to identifying themselves solely as masters (always dwelling in the grandiose or idealized mode) will find it difficult to accept an inferior role, and those prone to identifying themselves as inferior (and worshipful of their superior partners) will have difficulty tolerating feelings of superiority. However, by playing this game again and again, and dealing with the uncomfortable feelings that arise, the reversal of roles will get easier, sparking a fresher, more equal, more emotional relationship.

Game 5: Mirror, Mirror

Players: Husband and wife.
Activists: Both.
Setting: Home or hotel.
Aim: Get couple in touch with unconscious narcissism.
Game Plan: This game is best if the couple can put a mirror somewhere near their bed, on the headboard, on the ceiling, at the foot of the bed; or they can check into a hotel that features such mirrors.

The man and woman get undressed and lie across the bed, their arms around each other, and look at themselves in the mirror. Then they ask:

"Mirror, mirror, on the wall, who's the fairest couple of them all?"

They begin foreplay, fondling one another, and again turn to the mirror.

"Mirror, mirror, on the wall, who's the fairest couple of them all?"

They begin making love, and in the midst of it turn to the mirror again.

"Mirror, mirror, on the wall, who's the fairest couple of them all?"

Now they should turn on a recorded tape of their own voices, which repeats over and over, "You are the fairest couple of all! You, you, *you*! You are the fairest couple of them all! You and only you! No other couple will do! You and you and *you*!" This recording of their voices continues to play as they make love.

At first the couple may find this game fun, but after a while the repeated message and the image of themselves in the mirror quite likely will begin to grate and arouse other feelings. The game may then seem silly, and they may even want to stop. It is hoped that their motivation, fueled by a desire to achieve a better sex life and a better relationship, will inspire them to suspend judgment and see the game through. They may also become embarrassed, giggle uncomfortably, or get in touch with sadness or anger. Yet by the time they tire of looking at the mirror and shut off the recording, they will have reached a higher plateau of relating, being imbued with the realization that it does not matter whether or not they are the fairest couple in the world—only that they truly love one another.

7

Games for Obsessive-Compulsive Couples

One of the best known obsessive-compulsive couples is depicted in *The Odd Couple*, a play by Neil Simon, which was also made into a movie and then a television series. The story concerned a relationship between two men—one meticulously neat and controlling, the other a careless slob. Many male-female couples like this form "odd" relationships that often seem to have an undertone of strife.

Actually, the control freak and the slob are two sides of the same coin. Both have the same compulsions to be neat and controlling, but one acts them out while the other just gives up, assuming he will never be able to maintain such high standards. One is compulsively rigid, one compulsively loose. Both use their defensive posture to express anger—the one through manipulating and dominating the other with his neatness, orderliness, and stinginess, and the other through defiance, stubbornness, and slovenliness.

In *The Odd Couple*, Felix (the control freak) is always chastising Oscar (the slob), telling him to pick up after himself, wash his dishes, clean his room, eat better foods (not junk foods), and generally act more mannerly. Felix not only gets to order Oscar around, he also continually assumes and expresses moral outrage at Oscar's slovenly ways. Oscar, meanwhile, continually foils Felix's attempts to control him and shrugs off his companion's moral outrage. If Felix picks up after Oscar and gives him a lecture, Oscar almost immediately—and not always absentmindedly—drops something else. Hence he defeats all of Felix's attempts to control him and ends up infuriating his tormentor. In fact, the two are playing out the role of the scolding parent and the recalcitrant child.

This ritual goes on and on in the play, as it does in the real-life relationships of many couples, without resolution, and leads to constant discord that sometimes drives one (or both) into a frenzy. With respect to couples who are lovers, it even pervades the sexual arena: Control freaks want "managed sex," keeping it immaculate, keeping it tame, keeping it from getting out of control, robbing it of passion. They manage it by refusing to surrender to orgasm (which is regarded as loss of self-control), keeping their feelings under check by numbing their body, and insisting on washing their hands (not to mention other parts of their bodies) several times before and after having sex. They cannot stand for the seeds or juices that evolve during reproduction to linger on their bodies for long, nor for the smell to perfume themselves, their lovers, or the bedroom. On the other hand, slobs want "sabotaged sex." Acting defiantly slovenly and uncooperative in bed, they defeat their partner through sloth. They revel in the seeds and juices, wallow in the sweat, and delight in the smell of sex—not in a childlike, healthy way but in a gloating, provocative fashion.

Not all obsessive couples start out as opposite sides of the coin; sometimes two obsessively neat people—or two compulsively indolent ones—will team up. But when this happens, generally one or the other gradually takes the opposite role. The reason for this shift is that the obsessive perfectionist needs a manifestation of compulsive sloth as a foil, and the sloth similarly requires a demonstration of tyrannical neatness. If initially half of this dyad is missing, one of them must convert to fill that role. Each then comes to represent an aspect of one's own personality that is despised, disowned, and projected onto the other—where it can be soundly defeated.

Sex-therapy games for such couples must confront both their obsessive and their compulsive attitudes. These attitudes have strong secondary gratifications: The obsessive control freak gets to set the standards for the couple and to dominate according to those standards, while the compulsive slob defies those standards and undermines the obsessive person. Each derives immense satisfaction from his or her role, which he or she has played since early childhood. Old habits dic hard.

The following games address these secondary gratifications and attempt to loosen the grips of obsession-compulsion.

Game 1: Scientific Simultaneous Orgasm

Players: Husband, wife, and authority figure.
Activists: Husband and wife.
Setting: Home or resort.
Aim: Exaggerate orderliness in the sexual realm so as to underline the lunacy of such orderliness.
Game Plan: In this game there is a third player—an authority figure (or if necessary a book or cassette) serving to instruct the couple on the virtues of having a simultaneous orgasm. A therapist, physician, or friend can play the authority, or the couple can use a book—perhaps even this book—to serve in that capacity. Basically, what the authority does is encourage husband and wife to attempt simultaneous orgasms, suggesting that they are the key to both a better sex life and an improved relationship. As a matter of fact, if they play this game and *do* attempt to have simultaneous orgasms, they *will* achieve those goals. So the therapist, doctor, or friend can actually meet with the couple and confidently say something like:

"I recommend that the two of you try to have a simultaneous orgasm. This may take a while—and, given the state of your present sex life, it may seem impossible. In fact, it may seem impossible for either of you to achieve orgasms, period. However, if you work at it in a scientific way, you can accomplish it. Each time you have sex, you should each pace yourself and keep your sexuality under control, so that neither of you races ahead of the other. You can do this by each attaching an arm monitor that constantly checks your pulse and blood pressure.

"If one partner notices the other racing ahead, the partner who is behind should do something to distract the offender. The former can call out the other's name in a pronounced fashion; bark like a dog; make an idiotic face; squeeze the other's testicles or breasts to a point where that becomes annoying; slap the other on top of the head; talk about how some cultures eat snails; or ask whether or not the New York Mets will win the World Series.

"Practice this procedure every day at the same time if possible, and keep a chart of the pulse-rate and blood-pressure readings of each partner, the time orgasms occurred and their duration. Eventually, perhaps only after many hours of practice, you will be able to achieve a simultaneous orgasm—and along the way you will find that you will have to communicate about sex as you have never communicated before, because the teamwork entailed in achieving a simultaneous orgasm requires such back-and-forth. You will have to discuss everything that excites you and disgusts you, both sexually and in your relationship in general. Because everything that happens between you in and outside the bedroom—every thought, every feeling, every impulse, that arises concerning each other—affects your sexuality. Any questions?"

The couple either stays home or "runs away" to somewhere (a resort, a cruise, a vacation retreat) and follows the authority's orders. Whether or not the husband and wife actually achieve a simultaneous orgasm is not really important— the crucial factor is that they are scientifically making an attempt to do so, and in this attempt will be doing exactly what they long to do at the deepest level: keep their sex life under control. And now they have been given permission—in fact, ordered—to do so by an authority.

Being ordered to do something which they had previously done obsessively and compulsively will take the "sting" out of their obsessive-compulsive ritual. It will be rendered meaningless. Its chief purpose is, after all, to defend against both the loss of control and control by others.

In this game, control by another is de facto. So, in their quest for a "scientific" simultaneous orgasm, they will find themselves doing what they thought they wanted to do but will not be enjoying it. This brings up the feelings that had been previously warded off by their defensive posture—fears of being dirty, angry, rotten, shameful, getting completely out of control, going crazy, or being driven crazy. Encountering these ideas and feelings head-on can lead to another (and higher) level of relating and sexuality.

Game 2: The Nude Cleaning Lady/Man

Players: Cleaning lady/man and spouse.
Activist: Cleaning lady/man.
Setting: Home.
Aim: To provoke and seduce the obsessively neat partner out of the compulsive neatness mode and into the sphere of sexual play. To counter the fear of neatness of the slob-spouse.
Game Plan: The spouse who is the slob becomes the activist of this game, surprising the obsessively controlling partner at some appropriate time—perhaps on the weekend when the controlling spouse usually does the cleaning. On this particular day, the slob-spouse, well before the neat spouse has even awakened, begins scrubbing the floors, dusting the tables, washing the walls, polishing the furniture, sponging the windows, redoing the dishes, sanitizing the toilet, and even wiping off the light bulbs—all while in the buff.

The neat spouse wakes up, views this nude spouse-turned-cleaning-pro with incredulity, and probably exclaims, "What are you doing?"

"What does it look like I'm doing? I'm cleaning the house. I'm giving it the thorough cleaning it needs. I've finally understood that if I want it to be done thoroughly, I'll have to do it myself."

"Have you gone crazy?"

"No, I've finally become sane and realized that the house has to be kept in a certain order. You've been right to clean it and arrange it as often as you do, but you haven't gone far enough. It needs much more."

"Oh, really?"

"Yes. What have you been lying around for? Get out of bed and give me a hand—the baseboards are filthy!"

"I just waxed and polished them yesterday."

"I know, but there are still coffee stains on them. You need to really scrub them. Use the chlorinated Ajax."

"May I ask why you're in the nude?"

"That's part of the cleaning process. I need to be in the purest, most natural state in order to convene with the natural state of the world and of our house. Cleanliness is next to virginity."

"And what is the natural state of the world?"

"The natural state of the world? My dear, the natural state of the world is one of disorder and filth. Only through hard work and constant vigilance can we overcome this disorder and filth. Now please get up and give me a hand."

The neat spouse will most probably be amused and beguiled by this surprising turnaround. He or she may pop right out of bed and grab the slob-spouse and begin kissing or mounting behavior. However, the slob-spouse should resist this initial attempt.

"Don't. We have work to do."

"It can wait."

"How can you say that? You who have always been so gung-ho about everything? Look at the dust on those window sills! Look at the scum on those beams! Please—we have work to do."

"Just one kiss."

"I said no. Not until we finish."

The slob-spouse must insist on finishing all the work before he or she is willing to give in to the sexual desire. This, of course, mirrors the way in which the neat spouse generally operates—and it may well infuriate the latter. But this is as it should be: The anger that is generally acted out through the obsessive-compulsive rituals of orderliness will come to the surface and be dealt with. Both partners will be able to express their anger. The discussion—or argument—will at first center on the importance of orderliness as opposed to slovenliness, but eventually it will go deeper, to fears of inner disorder, filth, shame, loss of control, and insanity. These are fears that may lead to memories of past traumas.

The couple may or may not have sex on this occasion. When they do, it will have new meaning as a result of this game, and will be less hampered by obsessive-compulsive rituals.

Game 3: Mud Wrestling

Players: Husband and wife.
Activists: Both.
Setting: Home or away.

Aim: To appeal to slob's defensive fantasies and pull control freak down from immaculate pedestal, while regressing both of them to their childhood point of fixation.

Game Plan: The easiest way to do this would be to find a place where mud wrestling is offered as an amusement. A few resorts for couples provide it along with other "sports." However, if there is no such resort nearby, a bathtub can be turned into a homemade mud pit, or an isolated creek with muddy or sandy banks can be used as a wrestling arena.

There are several ways to convert a bathtub into a mud pit, depending on your taste. One option is to go to a novelty store and purchase enough boxes of safe-for-children "green slime" or "gunk" to satisfactorily fill the tub (this stuff is generally soluble in water, so easily washes away). Another is to sufficiently fill the tub with hot water and then stir in numerous boxes of jello or pudding until it thickens. (This will be more of a cleaning problem when you are finished, but that is fine, because it will become part of the game.) Yet another is to prepare the tub with water—or, if you can afford it, milk—and stir in chocolate mix. Still another is to crack and toss a dozen or so eggs into the tub water—if they're rotten, so much the better. Finally, one can simply add real mud from one's backyard to the water-laden tub. (This would also entail a lengthy cleanup afterwards, since the mud cannot simply be drained away—it might stop up the plumbing—but rather must be scooped up after the water has been drained out.)

If you have access to an isolated creek with a mud- or sandbank, you may wish to go there to wrestle. This is a bit risky, however, since you do not know what may be lurking in the earth (i.e., bugs, salamanders, frogs, snakes, poison oak or ivy), not to mention the water. So exercise caution!

There is no set procedure for this game. Each participant can attempt to pin the other in the mud for a three-count (as is done in regulation wrestling), or the couple can have other goals, such as trying to smear one another's face and other body parts. However it is done, the mere fact of indulging in such a game—of soiling oneself in this way—will break barriers and arouse new feelings. First, the game will be a regressive experience, harkening back to the age at which children love to play in mud puddles and sully themselves. Generally, obsessive-compulsives have fixations at this stage, for they usually had parents who were obsessive-compulsives and chastised the kids severely any time they got very dirty. This also corresponds to the potty-training (anal) stage, when children like to play with their feces and are scolded by parents for that. Second, the game gives the compulsive slob more than he or she bargained for, and challenges the rigidity of the obsessive neatnick. This experience creates stress and also liberates them from their guilt-ridden defensive attitudes. When two people are sitting in a tub with (the perfect example) egg on their faces, all pretenses quickly fall aside.

One thing usually leads to another, and the wrestling turns to erotic play and then to sex. It also leads to a fresh look at one's general modes of relating and of sexuality.

Game 4: Immaculate Consummation

Players: Husband and wife.
Activist: Both.
Setting: Hotel.
Aim: By playing out the "scolding parent" and "naughty child" scene in which hygiene is carried to an extreme, unconscious fears and resistances are traced to their roots and thus confronted.

Game Plan: The husband and wife arrange to go away for a second honeymoon. On their first night, before they go to bed, they must give each other a thorough scrubbing so that they will both be immaculate before engaging in sex. They take turns being the scrubber and scrubbee.

The scrubee stands in the bathtub or shower and the scrubber takes a cloth or sponge, soaps it up, and heartily rubs every spot on the other's body.

"Oh, what a dirty little boy you are," the wife should say to the husband as she scrubs him. ("Oh, what a dirty little girl you are," the husband says to the wife.)

The scrubbee takes extra pains in scrubbing the genitalia and other orifices.

"You're so dirty down there. I'm going to have to give you a good scrubbing."

"Yes, Mommy." ("Yes, Daddy.")

"How filthy you are down there."

"I am not!"

"Yes, you are."

The scrubbing should last a long time—perhaps an hour for each partner. Then a second scrubbing may be in order, just to take care of any filth that may have accumulated while scrubbing the other's body. After the cleansing, they take pains to put underarm and hygienic spray in appropriate places, to brush the teeth and wash out the mouth, and to powder the entire body. Finally, when they are both satisfied that they are immaculate, they go into the bedroom.

The bed has been made up with sheets and pillowcases that have been specially prepared for the occasion. Perhaps they have even been washed in boiling purified water and dried in the sun.

On the bed are two sets of transparent rubber gloves, such as a surgeon's, two masks of the kind doctors and nurses wear, and both male and female condoms. After they have put on their respective gear, they climb into bed. All through foreplay—during which they kiss one another with their masked mouths and fondle one another with their rubber-gloved hands—they talk about how clean they are.

"Oh, what clean sex this is!"

"Yes, yes—the cleanest sex I've ever had."

"This is how our first honeymoon should have been, instead of that sleazy affair we had."

"Neither of us had bathed for hours."

"It was horrible."

"Neither of us had used mouthwash."

"It was terrible."

"Or underarm deodorant."

"It was awful."

"Is your condom ready?"

"Yes, dear. Is yours?"

They proceed to have their immaculate consummation. By this time (if not before), the absurdity of the situation will have dawned on them. Like the preceding game, this one is a regressive experience, harkening back to potty-training days and to the days when Mom or Dad did the scrubbing. So the event will have many layers of meaning—the present will become mixed with memories of their honeymoon and each of their childhoods. This ridiculous experience will allow each to question, perhaps for the first time, the ritual of obsession and compulsion that has been dominating both their sex life and their overall relationship—and to begin to move on.

Game 5: Detailed Medical Examination

Players: Doctor and patient.
Activists: Both.
Setting: "Doctor's office" in the home.
Aim: Appeal to the fantasy of "playing doctor," which children of a certain age act out and which represents a natural curiosity about sexual details. Obsessives are big on details, so this game is designed to take the "detail orientation," eroticize it, and transform it into something more meaningful.
Game Plan: In this game, the husband and wife take turns playing doctor and patient. If they are an "odd couple," the control freak should play the doctor first, because he or she will relish that role more than the other, whereas the slob will relish the role of patient more.

At the appointed time the doctor comes out of his or her office (a converted bedroom, den, or basement will do) and asks the patient to come in. The patient does, and the doctor asks the visitor to get undressed and lie on the examining table.

"Now, this isn't going to hurt. I'm going to give you a thorough physical examination. Please bear with me," says the doctor.

The doctor then gives the patient a detailed examination, taking particular pains in examining the genital area. This examination may go on for an hour and will entail much poking, prodding, and tapping, as well as a lot of close inspection with eyes just inches from the part of the anatomy that is being examined—and sometimes involving the use of a magnifying glass, a hot sponge, or tools. (The doctor should use his or her imagination here, using tools—as well as methods—that will appeal most to the patient.)

When he or she has finished the examination, the doctor says, "Now, I'm going to have to take some sperm from you, in order to have it checked in the laboratory. (Or, "I'm going to have to arouse you in order to check your acidity level.") The doctor proceeds to bring the patient to orgasm, using a rubber-gloved hand and checking a watch while doing so. Afterward, he or she puts the sperm or vaginal fluid into a vial and labels it. (The examination may have other variations, according to the "tastes" of the husband and wife.)

"Now, that didn't hurt, did it?" the doctor asks when the examination is completed.

"No, not at all, doctor."

"Everything seems to be in order, physically speaking."

"That's nice, doctor."

"It's time for the postexamination interview, to analyze your thoughts and feelings about sex in order to determine if there is any psychological problem."

"Yes, doctor."

In the postexamination interview, the doctor probes the patient with regard to sexual problems. Naturally, these problems will involve the "doctor" (the spouse), so the answers will be tinged with irony. This irony, plus the heightened erotic sensitivity aroused by the examination, should lead to new insights into their sexual relationship, including the problems that surround it, and certain absurdities of its ritualistic aspect. It may also suggest possible ways out of their conundrum.

8

Games for Impulsive Couples

Dennis and Diane were a typical impulsive couple. They met, fell in love, and got married within two weeks. A few months later they began to lose interest in one another yet managed to keep their marriage afloat by constantly changing the external parts of their lives. They went from one temporary job to another. They lived from paycheck to paycheck and built up huge debts on their credit cards. They bought items they did not really need and took trips to exotic places that they did not have to visit. They were always getting "high" on one kind of drug or another. Their motto: "Live for today. Who knows what will happen tomorrow?"

Impulsive personalities cannot delay gratification—they must have what they desire *now*. They are fixated in what Freud called the oral stage. They need constant feeding and continual satisfaction of their cravings, and (like suckling infants) must have it immediately or throw a fit. They usually have had parents who did everything for them and did not allow them to learn to do things for themselves. Such parents typically say, "I don't want my child to go through the hardships I faced when I was growing up." So they go overboard and make it too easy for their youngster. The child becomes spoiled, unable to fend for himself or herself or cope with the daily grind of reality, commitment, long-range planning, and responsibility.

In many ways, the impulsive is the opposite of the obsessive-compulsive. While the latter is the product of harsh, overdemanding parenting and is therefore rigid and guilt-ridden, the impulsive person is the product of overly lenient, pampering parenting and is fickle and full of shame and rage. The obsessive strives to avoid feeling guilt by being super-responsible, and the impulsive tries to avoid rage by shunning responsibility and always seeking pleasure while living on the run. They are often addictive personalities—the drinker, the gambler, the overeater.

Sometimes impulsive couples are like Dennis and Diane, an alliance of two impulsives. At other times, impulsives can represent two sides of the same coin, much like obsessive-compulsives. In a typical alcoholic couple, for example, one spouse will be the *positive* impulsive—the alcoholic who needs always to assuage inner pain by binges of drinking, carousing, spending, and the like. The other spouse will be the *negative*—the crusader. Often this person is either a former alcoholic (A.A. is full of crusaders) or the child of an alcoholic.

The crusader and the alcoholic go through a ritual in which the alcoholic goes on a binge and the crusader chastises and condemns it. Then the alcoholic confesses his or her "sins" and begs for forgiveness, promising never to do it again, and the crusader forgives "one more time." Then the alcoholic goes on another binge (and so on). This ritual can be virtually endless, since from it each derives a secondary gratification from his or her role in the ritual. The alcoholic gets to relieve shame through frequent confessions and to act out rage through defiantly going on binges. The crusader gets to be morally superior and to prove that he or she is beyond such impulsive behavior. They both defend against and defeat a part of themselves that they wish to disown.

Impulsives use sex the way they use drugs—to avoid pain. Long-term relationships of commitment and responsibility do not appeal to them. Falling in love again and again, or constantly having new and different sexual experiences, does. It is no wonder, then, that Dennis and Diane had become bored with one another after a few months. They tried new sexual positions, having sex in novel places, swapping with other couples, and—finally—having separate affairs. Eventually they began discussing divorce. That is when they finally sought treatment.

I used three of these games with them, all of which at first they found hard to take seriously. Their resistance to doing anything that would interfere with immediate gratification made them find excuses for not playing these games or for doing them only halfheartedly. The game that finally got them involved was "One-Night Stand," since it appealed directly to their impulsive fantasies and offered the immediate gratification of acting out this fantasy. Once they had successfully played this game, they began to understand how they were blocking feelings and how those blocked feelings were causing them to need constant sexual novelty. Other games got them even further in touch with those feelings.

Eventually the games propelled them smack into the faces of issues they had been avoiding. After only a few weeks of the games, in fact, their therapy had reached a significantly deeper plane. After only several months of therapy, they decided to stay in their marriage and continue to work on their communication with each other rather than seek satisfaction outside of their union.

Game 1: Delayed Gratification

Players: Husband and wife.
Activists: Both.
Setting: Home or hotel.
Aim: To force impulsive couple to delay their sexual gratification in order to put them in touch with the feelings they are trying to avoid and help them learn to tolerate those feelings.
Game Plan: The rationale for this game is that as long as the members of an impulsive couple succeed in avoiding their deeper feelings of shame and rage, they will remain on a superficial, driven mode of existence, slaves of their own whims. This game uses sexuality to help them to both develop a toleration for frustration and delay gratification.

Basically, the couple is required to abstain from sexual intercourse (and indeed from *any* kind of orgasmic release, including mutual or self-masturbation) for a period of time—say one month. During this time the husband and wife are asked, instead, to get undressed several times a week and sit facing one another on a bed or couch and make love to one another only with their eyes and words—but without touch-ing. They should sit close enough to touch so as to have the opportunity to resist temptation. Then they should take turns making loving statements.

"I love your breasts," the husband might say. "I'd really like to squeeze them now."

"I love your big shoulders," the wife might say. "I'd really like to kiss them."

"I'd like to slide my hand down between your legs."

"I'd like to grab your buttocks."

"Your lips look so red and moist."

"Your eyes are glowing."

This verbal and visual lovemaking can go on for however long the couple wants it to. Perhaps the most difficult thing they have ever attempted will be to sit there naked and not give in to their impulses. In fact, the first time or two they try the exercise, they may weaken and fall into each other's arms and ravish each other. However, they should not then be fooled into thinking that the game is over. They should keep trying until they succeed in delaying their gratification for at least one session. And then they should continue to do so until they succeed in delaying it for a month—or some other agreed-on time period.

By delaying their gratification and withstanding tempta-tion, this couple are building up their egos—that is, their capa-bility to deal with the frustrations of life, of which the sexual frustrations of their marriage are merely a part. The exercise will also ameliorate symptoms of sexual dysfunction that are a part of impulsive personalities, such as premature ejaculation, shallow orgasms, and frigidity. In addition, as they resist temp-tation, they will get in touch with feelings of shame and rage.

The second part of this game consists of verbalizing thoughts and feelings that emanate from the frustration of their immediate desire.

"I hate waiting for sex."

"I don't see the point of this. It's stupid."

"Nobody ever waited for anything in my family."

"Nor in mine. Well, maybe we'll appreciate it more if we wait."

"I doubt it."

"You know, there's something strangely liberating about waiting."

"I know what you mean. We can stop running now."

Actually, they do end up appreciating sex more, as well as respecting themselves more. Eventually they get in touch with the shame that is associated with their inability to tolerate frustration, as well as with narcissistic rage and an attitude of entitlement.

Verbalizing these feelings and understanding them will put them in a different mode of being, one based more on mature self- and mutual respect. As they understand these feelings more, they will have less of a need to avoid them—hence their sexual relationship, and very likely their relationship in general, will be ever deeper and more rewarding.

Game 2: Three Sexual Wishes

Players: Husband and wife.

Activist: Either.

Setting: Home or hotel.

Aim: To appeal to the impulsive fantasy of instant gratification in an exaggerated way so as to underscore the tyranny of this fantasy.

Game Plan: This is a variation of the children's game "Three Wishes," played out in the sexual arena. In this version, one spouse—with *or* without the other spouse's prior knowledge and participation—asks the big question either before, or in the middle of, foreplay. They might be lolling in bed on a Saturday night, lush music about them, pink lights above them, the scent of incense in the air, when one of them sits up.

"If you could have any three sexual wishes that I could grant you right now, what would they be?"

"Any three wishes?"

"Yes."

"Hmmm. Seriously?"

"Seriously."

"Any three wishes? Any three at all?"

"That's right."

"What's the catch?"

"The catch is that before I grant you three wishes, you have to give me three answers."

"Three answers. What are the questions?"

"I'll tell you, but first you have to agree to the game. And once you make a commitment to the game, you can't back out, no matter how hard the questions are."

"Oh, so that's how it is."

"That's how it is."

"But then you have to keep your end of the bargain and grant the three wishes, no matter how hard they are."

"I will."

Once the respondent spouse agrees to the game, the activist spouse finds out what the three wishes are, promises to do them, and then asks three pertinent questions. These questions should be designed to get the truth out into the open so that it is not acted out sexually or in any other way. Examples: "Do you still love me?" "Are you bored with me sexually?" "Are you having an affair?" "Why don't you look at me when we're having sex?" These questions will be just as hard to ask as they are to answer, and are the very issues that are being avoided. Hence it will be quite difficult for both partners to play this game. It will entail much soul-searching, confronting, working through feelings, and give-and-take.

By the time the three questions are asked and answered, the first spouse may no longer be in the mood to have three wishes granted—at least not at that moment. So what might have been instant gratification of the wishes turns out to be delayed gratification. Inadvertently, the "active" impulsive has learned a lesson in tolerating frustration, taking responsibility, and long-range planning.

Game 3: Alcoholic Reversal

Players: Alcoholic (or gambler, or drug addict) and crusader.
Activist: Crusader.
Setting: Home or hotel.
Aim: To use reversal of roles to make impulsive alcoholic spouse aware of repeating pattern.
Game Plan: Following the alcoholic's latest binge, that

spouse will (as usual) confess, apologize, and promise not to drink or misbehave again. The crusader spouse will as usual accept the apology. However, this is as far as the repeating pattern will go. When they go to bed and the alcoholic spouse is about to make love with the crusader, clinging to the latter like a shameful child, eyes begging for forgiveness, the crusader suddenly begins mirroring the alcoholic's behavior.

As they lie in bed, the crusader turns to the alcoholic and says, "I too have a confession to make."

"What's that?"

"I've been acting like a moralistic pig, and I'm sorry."

"I don't understand."

The alcoholic will probably be stunned and look aghast at the crusader. This is not what the alcoholic expects or wants. What is expected and wanted is one rebuke on top of another, all of which will serve to assuage guilt and justify future defiance. The crusader continues to break the pattern: "I'm really sorry. Please forgive me. I promise I'll never do it again."

"Do what?"

"Act like such a moralistic pig. I know I've caused you a lot of pain and misery with my moralistic tyranny. Yes, it's true—I've been getting furious at you for drinking and being a self-righteous tyrant about it. And because of my tyrannical attitude I've driven you to drink again and again. Please, please say you'll forgive me. I promise I'll never, ever do it again." The crusader should go on in this vein until the alcoholic is either convinced or begins to protest.

Of course, the crusader's speech must be convincingly given. This will be the tricky part, since the crusader's own defensive attitude is that he or she is innocent, and that the only problem in their relationship is the alcoholic. In fact, the crusader's very identification and ideal image depend on seeing things just this way—so, giving up this posture in order to play this role will be quite a stretch as well as a learning experience. (Even if the crusader is not quite convinced of the truth of the speech in the beginning, it may well ring true as it is delivered.)

Meanwhile, the alcoholic will be stunned by this reversal and confused to see the crusader asking for forgiveness rather than preaching. This confusion is a good confusion that will lead to insight. Both the alcoholic and the crusader will be enabled to see the folly of their standard roles.

The fact that this reversal takes place in bed, during the middle of foreplay, adds an erotic emphasis to the proceedings. We are seldom more vulnerable than when we are in bed, naked and locked in sexual embrace:

Game 4: Responsible Sex

Players: Husband and wife.
Activists: Both.
Setting: Home.
Aim: To integrate responsibility and sexuality.
Game Plan: This is a simple exercise that a husband and wife can practice regularly several times a week, just as some people go to a gym and work out. Only this is a lot more fun.

The husband and wife go to their bedroom and, after their usual foreplay, get into a position of sexual intercourse in which they are face-to-face. This could be one in which he is sitting on an armless chair and she is straddling him, or they are lying side-by-side, or they are in the standard "missionary" position. When they are in place and have begun the act of intercourse, they should look into each other's eyes and say something like the following:

He: "I'm having sex with you, and I take responsibility for having sex with you."

She: "I'm having sex with you, and I take responsibility for having sex with you."

He: "I'm looking at you, and I'm taking responsibility for my feelings."

She: "I'm looking at you, and I'm taking responsibility for my thoughts."

He: "If I want the sex to work, I have to take responsibility for making it work."

She: "If I want the sex to work, I have to take responsibility for giving myself to the experience."

He: "I see you and I'm having sex with you."

She: "I see you and I know you're inside me."

He: "My eyes want to stop looking at you because I don't want to take responsibility for the sex."

She: "My eyes want to stop looking at you because I don't want to commit myself to you."

He: "If I really love you and care about you, I will take responsibility for each sexual experience."

She: "If I really love you and care about you, I will take responsibility for my own participation."

He: "If I really love you and care about you, I will look at you and understand that sex is not just for the moment but is an act of reproduction that is eternal and may produce off-spring."

She: "If I really love you and care about you, I will look at you and really be with you while we're having sex."

He: "I'm taking responsibility for my sexual feelings."

She: "I'm taking responsibility for my sexual feelings."

These and similar phrases should be spoken like a litany, in a low voice as the sexual experience goes on. There will be a tendency to want to laugh, to break eye contact, to pull away from the litany and just have sex. The couple must fight against this impulse and stay with the exercise. If one or the other partner feels like laughing, that is all right. Laugh it all out, and then continue with the exercise. If one member begins blinking or looking away, that is all right, too. Blink as much as you want, look away as long as you want, but then continue the exercise.

Of course, as the exercise continues (and particularly when it is repeated regularly at the same time several nights a week), it will bring about a new sense of connectedness between the man and woman, one they have been avoiding. It will also make clear the reasons why they have been avoiding it. The sense of connectedness will arouse the fears of responsibility and commitment that lurk beneath their impulsive behavior. However, by acting out these fears and avoiding such feelings, they are preventing themselves from enjoying the deepest wellsprings of their sexuality.

When they do experience such feelings, they may add more statements to the litany, such as: "Now I'm looking at you and feeling afraid to make an honest commitment to you." Or: "Now I don't want to look at you because you may expect things of me if I do."

Like doing any exercise over a period of time, the longer this game is played, the better the results.

Game 5: One-Night Stand

Players: Playboy (husband), playgirl (wife), and authority (perhaps).

Activists: Both spouses.

Setting: Hotel.

Aim: Play out a primary fantasy of impulsive couples, but with a twist: Instead of writhing and running, writhe and review.

Game Plan: This game can be played with or without an actual authority figure—a therapist or a nonprofessional friend. (This book may serve as the authority if no other is available.) The authority instructs the couple to go to a hotel or motel and play out their impulse to have a one-night stand—with each other. However, this time they are to end the sham affair different than usual. Usually, the morning after such an event is awkward as each partner finds the quickest way to the exit. This happens because in a one-night stand a couple quickly achieves a false intimacy based on their impetuous sexuality—but they have not really communicated and bonded so that they can sustain a relationship, and they are reluctant to do so. Hence, the next morning they find themselves having been sexually intimate but thinking all kinds of thoughts about one another that they do not want to talk about. Indeed, they fear that talking about these things would be harmful. (In reality, in many cases talking about these things would lead to a deeper, more realistic relationship.)

The couple is therefore instructed to go to a lodging for the weekend and "pick each other up" as if they were meeting for the first time. They are encouraged to dress like a single playgirl and playboy, whatever that label conjures up for them. The game might start in the hotel's bar on Saturday evening. The man comes into the bar and spots the woman sitting alone. He sallies forth, smiling confidently.

"Excuse me," he says. "Would you mind if I join you?"

"No, not at all."

"Thanks. May I buy you another drink?"

"Why not?"

"You know, I hope you don't take this the wrong way—but you really have beautiful eyes."

"Oh, thank you. I hope you don't take this the wrong way, but you have terrific buns."

"Oh, thank you."

"You know, you remind me of somebody. Somebody I once fell in love with."

"Oh, really? You know, now that I think of it, you also remind me of somebody I once fell in love with."

"Isn't that remarkable?"

"I'll say!"

"Cheers!"

"To your sexual health!"

The conversation continues in this vein—deliberately full of sexual innuendoes and ironic comments common to singles bars. They have a few drinks, and then the man asks the woman to go up to "his" room to see—well, his collection of silk ties, shall we say. They go there, and he continues to seduce her, and she continues to succumb—with varying degrees of resistance.

She: (As he kisses her.) "What are you doing?"

He: "I'm not sure."

She: "I think it's called kissing."

He: "That's right. Kissing. I knew there was a name for it."

She: "Did I say you could kiss me?"

He: "No. But you didn't say I couldn't."

She: "What are you doing with your hands?"

He: "I don't remember."

She: "But you're doing it right now. How can you not remember it?"

He: "I have a very bad memory sometimes."

She: "I don't know if I want to do this."

He: "I don't know either."

She: "I don't know if I'll respect myself in the morning."

He: "The morning's a long way off."

She: "What did you say your name was?"

He: "Masterson. Hank Masterson."

She: "Glad to meet you, Hank. What do you do, Hank?"

He: "I'm a ladies' underwear salesman. Which reminds me: What's your name—if I may be so bold?"

She: "Kitty."

He: "Kitty what?"

She: "Just Kitty."

He: "Oh, I see—the mysterious type. And what do you do, Kitty?"

She: "Well, you promise you won't laugh? I'm a stripper. I work in the Hellcat Lounge."

He: "So this is called kissing?"

She: "I think that's what it's called."

They continue to stay in character throughout the night, making love in both old and new ways. The test will come in the morning. Each may want to run, but instead they must stay in bed and spend an hour or so talking about the experience—how it made them feel to play out their roles, and what they did not say to each other last night. During this part of the game, each partner becomes the other's disciplinarian, preventing the other from running. The subsequent review will beneficially transform their sexuality.

9

Games for Perverse Couples

Games for perverse couples abound both in life and in literature. In days of yore, such games were played secretly in bedrooms, back alleys, and basements. Today, one encounters more and more couples engaging openly in perverse games, such as those advertised in the "adult classified" pages of newspapers. They are referred to as (S&M), role-playing, and sometimes even sex therapy!

In perverted sex, sexuality is turned into an experience whose aim is no longer primarily the expression of love and tenderness and/or the reproduction of the species. Rather, it is concerned with discharging aggression and assuaging envy, guilt, and fear, as well as with repeating compulsively certain sexual dramas that are rooted in unresolved childhood sexual experiences and conditioning. Children who are molested by an uncle or aunt often grow up to be pedophiles themselves. In molesting little boys or girls, they will not only reenact what was done to them but also will discharge repressed anger, displacing it onto their new sexual objects. A boy who is dressed in a girl's clothing by a mother or aunt who was disappointed to have a male child often will grow up to be a transvestite. By wearing women's clothing, he both reenacts his childhood conditioning and discharges aggression (by defying convention), castration fear (hiding behind a skirt), and envy of women. Children who are severely punished or humiliated for some kind of infantile sexual play will grow up to enjoy sadomasochistic sex. They will both reenact the abusive trauma of their childhood and discharge repressed aggression onto their new objects. (The favorite fantasy of couples who engage in S&M is for one of them to play the punishing parent and the other to play the naughty child.)

Perverse couples are among the least inclined to go to a therapist—for two reasons. First, their perversion was developed in secrecy, and they wish it to remain so. Second, they strongly believe that there is nothing wrong with their perversion, and fear that the therapist will tell them that there is. Indeed, today the whole topic of what is perverse and what is not has become controversial. Formerly, homosexuality was considered a perversion—but homosexuals protested and persuaded the psychiatric establishment to drop being gay from its list of perversions. Transvestites, who call themselves "cross-dressers," are now likewise waging a battle to have their sexuality considered normal. Fetishists—men (and in rare cases, women) whose primary sexual drive is attached to a symbolic object, such as shoes, gloves, underwear, or hats—are also convinced that their sexual orientation is different but still harmless and therefore should be accepted in the mainstream. Sadomasochists similarly feel that a little black leather, bondage, and humiliation by mutual consent should not be considered perverse. Even those whose sexuality is criminal in nature, such as men who molest boys, defend what they do. (I have even heard of an organization formed to propagate the benefits of an older man's befriending and having sexual relations with a boy.)

In order to avoid getting into the controversy about the shifting definition of perversity, I refer here only to sexuality in which symbolism predominates over authenticity and in which hate, fear, humiliation, envy, and guilt (rather than love and tenderness) permeate the sexual act. Hence I am writing about the sexual relations of perverse couples that are comprised of immature, symbolic reenactments of some parent–child trauma, and whose sex is replete with rituals of punishment, revenge, degradation, appeasement, submission, and defiance—not about here-and-now adult dyads between mature and independent people.

Perverse couples play sexual games all the time, but their games are not truly therapeutic. They become swingers, compulsively swapping partners with other couples; they cross-dress with one another and with other couples; they play dominance and submission games, using such various paraphernalia as black leather belts, whips, and chains; they have sex in public places, in trains, and in airplanes; they engage in indiscriminate orgies; and they will experiment with threesomes, and even with using dogs or farm animals as objects. These games are gratifying to many of their players in much the way that drinking a glass of whiskey may be gratifying—but they aren't therapeutic: Like whiskey, they have side effects and are ultimately toxic to the body.

The following games are variations of some of the common themes of perverse couples, but with twists designed to bring about insight, resolve perverse fixations, and facilitate a more profound, authentic form of sexuality and relating.

Game 1: Gender Reversal

Players: Man (wife) and woman (husband).

Activist/s: Either can surprise the other, or the game can be played mutually.

Setting: Home, or hotel.

Aim: To activate and resolve unconscious or conscious envy and fear of the opposite sex through gender reversal.

Game Plan: One evening, a spouse will come home to find the other spouse has changed genders—that is, has dressed in clothing of the opposite sex. The husband may come home to find the wife dressed in a suit and tie, wearing a "crew-cut" wig, sporting a mustache or beard, smoking a cigar. The wife may come home to find the husband wearing a dress, a long-haired wig, lipstick, rouge, and high-heeled shoes. In this case the element of surprise will add spice to the game. Or, the husband and wife may simply pick a night to mutually play the game.

In this case, the wife comes home to find her husband dressed in women's clothes. She proceeds to play the male role of being the sexual aggressor. If the game is mutual, she may wear men's clothes, too.

"Well, well—look who I've found," the man (wife) says to the woman (husband). "How pretty you look in your pink dress."

"Thank you. And you look quite handsome in your suit. And that mustache is quite manly."

"Thanks. I grew it yesterday. I think I'll just sit beside you."

"Why do you want to sit beside me?"

"In order to see your beautiful eyes up close and personal."

"Why do you want to see my beautiful eyes?"

"Because I may want to kiss them."

"I don't know if I want you to kiss them."

"Or perhaps I'll kiss your moist red lips."

"I don't know if I'd like that, either."

"How about *this*? Do you like my hand under your skirt?"

"I'm not sure."

"Hmmm. Look what I found under your skirt!"

"Stop that."

"That's interesting. You're a woman with a penis."

"Oh, yeah? And what's that in your pants? You're a man with a vagina. Wow!"

"There's nothing wrong with vaginas."

"There's nothing wrong with penises, either."

"What are you doing now?"

"I'm taking off your blouse."

"Did I say you could do that?"

"May I?"

"I suppose so."

"And how about your skirt. May I take it off, too?"

"All right. But be gentle."

"I will."

"And promise me something."

"Yes."

"When you go drinking with the boys, be kind."

"I promise."

The man takes off the woman's blouse, skirt, panties, bra, and stockings, and then undresses himself, engaging in more banter as he does so. When they begin to have sex, the man (wife) continues to be the aggressor and the woman (husband) the passive–seductive recipient. The man with a vagina may, for example, get on top and straddle the woman with a penis, thrusting himself against her member with manly force and passion. Or, he may lie on the bottom and wiggle his hips upwards into the woman. In general, he will be the man she always wanted to be, and she will be the woman he always wanted to be.

"Oh, what a man you are!"

"And what a woman you are!"

Afterwards, they should lie side by side and remove all the trappings of the gender reversal—the makeup, wigs, rings, and other items—and become themselves again.

Now comes the hard part of the game: talking about their feelings. As with previous games, this is the crucial step; without it, the game remains a shallow piece of acting-out. The couple should lie facing one another, and talk about what it felt like to be a man or a woman, what they wanted from each other, how they felt sexually, and what they remembered from the past—particularly their childhood. They should be candid about the negative feelings that came up. The man may say, "It felt kind of good not to be a man; not to have to initiate sex and risk rejection." The woman may say, "It felt strong to be a man—I could make all the moves and didn't have to restrain myself." He may say, "You know, I got in touch with how much I'd like to be a woman, and how much I resent women—and you." She may say, "I got in touch with how much I hate men and their arrogant attitudes about their penises."

This conversation can go in many directions. The important thing is to let the conversation continue to go wherever it will—no matter how embarrassing, risky, or seemingly insane.

Game 2: Tie Me Up

Players: Husband and wife.

Activists: Both.

Setting: Home or hotel.

Aim: Appeal to perverse fantasies in order to awaken repressed feelings and memories.

Game Plan: The perverse couple will already have played—or thought about playing—games like this. Indeed, they may have engaged in sadomasochistic sex to the point where they are saturated and cynical. To some, this game may seem mild, but if the instructions are followed precisely, it contains an important twist that can bring about change.

Basically, in this game one spouse ties up the other and pretends to degrade and ravish him or her. However, in this game the spouse who in actuality or fantasy generally prefers to be the dominant spouse is encouraged to take the role of the victim, and vice versa. When the spouse who wants to be dominant is asked to be submissive, and the spouse who wants to be submissive is asked to be dominant, each is forced to confront aspects of themselves that they unconsciously deny. The one who prefers being dominant disowns his or her potential "weakness" and the associated negative judgments of a degrading sort, projecting them onto the victim. The one who prefers being submissive feels guilty and seeks punishment, denying his or her rage and masochism yet identifying with the aggressor (the spouse) as the latter performs degrading and abusive acts.

The game is straightforward and is played by mutual consent. It begins with the "dominant" partner (in this example, the wife) ordering the "submissive" partner, "Take off your clothes and lie on the bed."

The submissive partner says, "Yes, Madam."

"And hurry up about it."

"Yes, Madam."

The submissive partner undresses but the dominant one remains dressed. When the submissive partner is naked and lying on the bed, the dominant one ties him, hand and foot, to the bedposts or railing (or in some other convenient way), using a soft rope that will not burn the skin. When the submissive is tied down, the dominant stands over the bed, grinning.

"Now you're going to get what you've had coming."

"I am?"

"You are. You've been bad, and you know you've been bad."

"I have?"

"Yes, very bad. I'll show you how bad you are. Do you like *this*?" (Puts hand on submissive partner's crotch.)

"Yes."

"You see how bad you are? You like my hand on your dirty thing."

"Yes."

"Yes, Madam!"

"Yes, Madam!"

"Now I'm going to punish you."

"What are you going to do?"

"I'm going to stick your dirty thing inside my dirty thing."

"Please don't do that."

"Sorry. You asked for it."

"No, no, please—spare me!

"Sorry. This is what a dirty nerd like you deserves."

The conversation will, of course, vary according to the whims and discretion of the participants. It must be emphasized that this game is to be played only by mutual consent, and if during any part of the game one or the other participant becomes upset, it should be stopped and the submissive partner untied. However, as long as the game is enjoyable (or, at minimum, endurable), it should continue. When the dominant initiates intercourse, he or she begins the second phase of the game.

"How do you feel now?"

"I like it."

"You like to be degraded, don't you?"

"Yes, Madam."

"You like to be humiliated."

"Yes, Madam."

"Because you're bad."

"Yes, Madam."

"Why are you so bad?"

"I don't know."

"What do you remember?"

"I don't remember anything."

"Yes, you do."

"I remember being bad."

"How were you bad?"

"I touched myself down there."

"Did you get caught?"

"My mother caught me. She said I was bad."

"What else?"

"My uncle touched me down there."

"And you liked it, didn't you?"

"Yes, sometimes. But I also hated it."

"You liked it. That's why you need it now."

"Maybe."

This line of dialogue is confrontational and can arouse disturbing feelings. By arousing these feelings in the context of sexual intercourse (that is, in the act that recreates a traumatic sexual situation), these disturbing feelings have greater access to the repressed memories beneath them. By getting in touch with these hidden memories, feelings can be worked through and will cease to be compulsive—hence the need to act them out will diminish. Instead, the memories will become stronger and will take precedence. Sometimes this happens immediately the first time the game is played, and the couple needs to stop the game and work through the feelings right away. But sometimes the game has to be played several times before this resolution occurs.

The working-through process can take hours, weeks, months—even years. As it evolves, the sexual relationship will change. Hence, when this same game is played a few months later, the conversation may well run like this:

"You like that, don't you? You like being degraded."

"No, not so much."

"No?"

"I'd rather you just kissed me. I don't need to be humiliated anymore. I'm not dirty. I'm not bad. Just kiss me."

"Kiss you?"

"Yes. Untie me and kiss me."

Eventually this game, like most of the other games in this book, becomes unnecessary, and the couple can proceed to the games in chapter 14.

Game 3: Look, Mom—I Have a Penis!

Players: Mother (wife) and son (husband). Variation: Father (husband) and daughter (wife).

Activists: Both.

Setting: Bathroom.

Aim: To appeal to voyeuristic and exhibitionistic fantasies while providing a reparative response to a childhood fixation.

Game Plan: The game can be between mother and son ("Look, Mom—I have a penis!") or father and daughter ("Look, Dad—I have a vagina!"). This present version is written for the former duo.

The "mother" starts out by giving the "son" a bath. She bathes parts of his body, taking extra time with the penis. The game is intended to harken back to the age of sexual discovery (between two-and-a-half and four)—when, researchers have determined, perverse forms of sexuality develop. Mothers and fathers at this stage are an all-important influence on the direction that a child's sexuality takes. The mother in this game starts by doing what many mothers do—focusing on the cleanliness of the child's sexual organ. (*Note:* When this game is between "father" and "daughter," the father bathes the daughter in the same way.) "You must always keep this clean, son," she says as she scrubs his member.

"Yes, Mom."

"Now, I've washed your back, your underarms, your rear, and your penis. *You* wash the rest."

"Yes, Mom."

She sits on a very high stool (or such), so as to give the impression that she is sitting high above him, just as a mother does. She may also be dressed in a mother's type of dress—an old-fashioned one, and wearing makeup and hair in a motherly way (in a bun) as well. She sits with arms folded, as if ready to judge and rebuke.

The son proceeds to play with himself as Mom watches.

"Look, Mom, I have a penis."

"So I see."

"It's a very special penis."

"Is it?"

"It is! It really is! Look at my penis."

"I'm looking."

"See what it can do? It can get big like this."

"That's a wonderful penis you've got."

"And it can swing from side to side like this."

"What a wonderful swing."

"And up and down, too."

"Oh, my dear—how amazing!"

"That's nothing. Wait until you see what it can do next!"

"What's that, I wonder?"

"You'll see."

Mom and son continue to chat, elaborating on this theme. The important thing is that the mother in this case provides a caring, empathic response to the boy's phallic exhibitionism rather than the inappropriate response that he most likely got from his own mother. Many mothers say, "Stop that—that's naughty!" or "If you do that, you'll go crazy later in life." They may just give a scowling or embarrassed look. Or they may smile flirtatiously. This mother does none of these things, neither encouraging nor discouraging his infantile sexual play, yet instilling positive judgments about his penis.

The son masturbates himself until he reaches orgasm, and Mom looks on without praise or curse.

"Did you see what my penis can do?"

"Yes, yes. It can do many things."

"Did you see the magic serum spurt out?"

"Yes, I saw it."

"Isn't it a wonderful penis?"

"Of course it is."

"You're a good Mom."

"Thank you."

"Better than my other Mom."

"I hope so."

"So you think it's a great penis?"

"It's the best. Number one."

The game ends at this point. No sexual contact is made between mother and son—not only because it would be "incestuous," but also because it would be detrimental. This is crucial. If contact is made, the game becomes another exercise in perverse sexuality for the sake of perversity and loses its impact. Again, the purpose of the game is to allow the voyeuristic and exhibitionistic tendencies to be played out while giving mother and son (or father and daughter) a chance to have a reparative experience. Following this experience, there should be a discussion of what happened and how it made each feel. The discussion should be held in the living room, with the principals fully clothed.

Game 4: Spin the Bottle

Players: Husband and wife.
Activists: Both.
Setting: Home or hotel.
Aim: Appeal to perverse fantasies of both partners in order to get them in touch with unconscious thoughts and feelings.
Game Plan: This is a variation of the game that adolescents play. In the original game of Spin the Bottle, a group of teenagers sits in a circle while one member spins a bottle. When it stops, the spinner of the bottle must kiss the person of the opposite sex sitting closest to where the head of the bottle points. In this version the man and wife each take turns spinning the bottle. If the bottle lands on the spouse, the spinner may ask the spouse to do something sexual that the latter has not done before. If the bottle lands on the spinner, the spinner must do something sexual that he or she has not done before, as requested by the spouse.

Actually, this game can be played not only by perverse couples but also by bored, depressed, narcissistic, obsessive–compulsive, impulsive, angry, or totally uninterested couples as well. Perverse couples, naturally, will get off on it best, since it gives them permission to be perverse. However, this game demands that even perverse couples find something sexual they have not done before—that is, stretch themselves beyond their sexual boundaries. By doing this, they are forced to confront the meaning of limitations in sex as well as the feelings that are associated with those boundaries (and with the ones that they have already broken). This raises them to a new awareness. Thus the game is fun, invigorating, and thought-provoking.

Like the other games in this section, it should be followed by a discussion geared to working through any feelings that arise. The game can be played again and again, with fresh results each time.

Game 5: Sweet, Wholesome Sex

Players: Husband and wife.
Activist/s: One or both.

Setting: Home or hotel.

Aim: By going against the grain of the perverse couple's fantasies this game arouses the feelings that are being acted out.

Game Plan: The last thing that a perverse couple wants is to have sweet, wholesome sex. They do not at all feel sweet or wholesome, and in fact are generally cynical. They are most likely to think that sweet, wholesome sex is for nerds and birds. They try to avoid straightforward sex because they need to ward off the feelings that such intimacy would create. This game is a paradoxical approach that "kills them with kindness" and brings up those feelings that are warded off by their perverse sexuality.

One evening, the activist spouse (with or without the cooperation of the other) sallies up to the more perverse spouse and says, "I have an idea. Let's go to bed and make love."

"Just like that?" the perverse spouse answers, surprised.

"Just like that."

"Maybe we should call up Jill and Bob and swing with them."

"Not tonight. I just want to make love with you—simply and deeply."

"Well, how about if we borrow Howard's German Shepherd?"

"No dogs."

"All right. Wait, I'll go and get out the whip."

"No whip."

"Then the handcuffs, at least."

"No handcuffs."

"What about the vibrators?"

"No vibrators."

"Dildoes?"

"No dildoes."

"Oh—*I* get it: You want to do some new role-playing?"

"No role-playing."

"I don't understand."

"I just want to make love to you, simply and deeply."

"Just you and me?"

"That's right."

"How about just a little safety pin in my nipple?"

"No pins."

"Boring!"

"Maybe."

Of course it will not be at all boring. Boredom, as we noted in the second chapter, is a feeling that covers up some other thought or feeling that is being avoided. The activist spouse leads the perverse spouse to bed and proceeds to make love in a very straightforward way, saying, "I love you," and "I appreciate you" and "You're wonderful" as he or she softly and tenderly kisses the other and engages in standard lovemaking.

The perverse spouse (or both) may indeed experience boredom at first:

"I'm sorry, but I just can't seem to get into this."

"How come?"

"I don't know. It's too . . . too . . . straight."

"What else are you feeling, other than boredom?"

"What else? I don't know. Maybe annoyance."

"What's that about?"

"That something is too . . . too . . . direct . . . too sweet."

"Tell me about it."

The perverse spouse talks, and then the activist spouse talks—and they will probably learn something new.

10

Games for Angry Couples

Many couples are angry some of the time, either overtly or covertly. However, some seem to be continually at odds with one another, are constantly battling it out. Such couples are not of any one character type, as are most of the couples discussed previously. They are of mixed types, combining hysterical, sadomasochistic, obsessive–compulsive, passive–aggressive, and narcissistic components—among others.

People tend to become addicted both to displays of anger and to the rituals surrounding them. These rituals serve to ward off feelings of emptiness, powerlessness, depression, or other kinds of frustration, while giving life some kind of twisted meaning. Generally these rituals involve power struggles over who is good and who is evil—that is, who is to blame for their bad relationship. Particularly angry relationships involve the same kind of warfare one sees between countries—marriage partners becoming enemies who demonize one another and commit acts of verbal and physical violence designed to defeat, if not destroy, one another.

One such couple who came into treatment consisted of an alcoholic man and his equally alcoholic wife. He was a failed musician who blamed his failure on his wife: "If it weren't for you, I could have been somebody." He held her responsible for holding him down, including in his rationalization her bringing their four children into the world. (His wife had become pregnant when they were only seeing one another, and he'd felt obliged to marry her.) He felt his life had been literally destroyed by his wife, and he continually lambasted her for that. He would typically come home drunk after work and begin hammering away at her, verbally and sometimes physically, taking out any daily frustration on her and always ending up by blaming her for everything.

The wife, meanwhile, was a martyr type who also blamed her spouse for everything: "I don't know why I put up with this hell," she would moan again and again, each time her husband battered her. She clutched her four children tightly to herself and turned them against their father, telling them he was an evil drunkard and that she should leave him. But when they gave their permission for her to do so (even urged her to), she could not do so. Down deep she was too insecure to separate from him because she had had a disturbing childhood, yet had never been able to separate from her psychologically abusive parents. Nor could she refrain from talking back to him, and sometimes instigating quarrels—both of which actions her children begged her to avoid.

Like so many battered and battering couples, these two were dependent on their nightly ritual of violence and expiation, caught up in the battle over who was wrong or right. Just as nations in battle cannot be dissuaded from their war aims by objective outside forces (e.g., the United Nations), so such couples typically cannot be steered away from their battle plans. Hence, they are difficult to treat in therapy.

One possible entry point is to go with the couple's anger and use it to put them in touch with their deeper feelings. (Anger is a secondary emotion.) I had this couple engage in a foam-bat fight in my office. (I had no trouble getting them to take this exercise seriously.) At first I let them just hit one another until they tired. Then I added sentences for them to say as they were hitting. Eventually, after a few weeks, they began to come out of their rage state, and to discuss and unravel their various characterological attitudes toward one another. It was extremely difficult for both to give up their outrage as well as their notion that each was the victim of the other. Many couples would rather be miserable and feel morally right than to feel the relief and happiness of taking responsibility for their end of a dysfunctional relationship.

After several of the office fights had ventilated their anger some, I assigned the Nude Foam-Bat Fight series to be held at home. *In toto*, the period in which they used this sexual game was brief. It was the first "swing" in the direction of their recovery as a couple.

Game 1: Nude Foam-Bat Fight

Players: Husband, wife, and therapist (optional).
Activists: Both.
Setting: Home.
Aim: Channel anger into a harmless physical battle that serves to elicit deeper feelings and bring about insight.
Game Plan: If the couple is in therapy, they may begin by having a foam-bat fight in the therapist's office. If not, a friend may play the role of therapist, using this book as a guide. Supervision is often required (just as athletic games require referees) in order to keep the rules as well as set the framework of the contest.

The rules of this game are that the contestants should stand facing one another in the middle of a large room in which all furniture is out of the way. A formal "ring" may be set up if desired, roped off like a boxing ring, with some kind of padded floor (a king-size mattress or thick rug will do). At a designated start time they should begin striking one another with foam bats, aiming above the waist and avoiding the face. Foam bats can be bought from toy stores, or made with 1" × 10" × 18" foam strips rolled up like newspapers.

In the beginning, the combatants will need close refereeing. After all, they are used to their usual no-holds-barred brawls. Once accustomed to the rules and able to abide by them, and upon their becoming tired of shouting the usual destructive putdowns at each other, they should be encouraged to say the following sentences as they flail away with the bats. The first fight should happen while they are fully clothed.

"You're the cause of all my misery."

"No, *you're* the cause of all *my* misery."

"I'm completely innocent and you're completely guilty."

"No, I'm completely innocent and you're completely guilty."

"I wish you were dead."

"And I wish you were dead."

"As long as I stay angry at you, I don't have to take responsibility for my own misery."

"And as long as I keep blaming you, I don't have to take responsibility for my own misery."

"Just think, we can keep trying to destroy each other like this for the rest of our lives."

"Yes, we can keep destroying each other—isn't it great?"

Each couple will embellish these statements with comments that fit their own particular situation. Once this stage of the game has been mastered, the couple is ready for the *Nude Foam-Bat Fight*. Basically, the same rules apply, and the couple is asked to say the same sentences as they hit one another's nude bodies.

The element of nudity adds two other dimensions to the game. First, it makes them much more vulnerable and hence heightens the impact of the encounter. (When we are naked, we cannot as easily pretend or hide behind our image.) Second, nudity is erotic and so provides a temptation to channel aggression into sexuality. Channeling aggression, of course, is one of the chief things that sexuality does throughout the animal kingdom. Thus, if this game is played well and with authenticity, eventually the couple will toss aside the bats and begin making love—and the lovemaking will have a different quality than ever before.

Caution: This game is not recommended for couples whose anger is out of control unless they are in couples therapy.

Game 2: Rose Petals

Players: Husband and wife.
Activist: Husband.
Setting: Bedroom.
Aim: Melt wife's anger by surprising her with an atypical, extremely romantic and erotic gesture.
Game Plan: This game is for couples in which the wife is ostensibly the angriest partner. She may, for example, be one of those types who is constantly enraged at her husband because she does not feel that he is ever truly tender or caring or sexually attentive—or because he is always working,

watching football, going out with the guys, and otherwise keeping her at an emotional distance.

On a designated evening or weekday afternoon, the wife comes home to find the husband dressed in a silk robe and holding out a dozen roses to her.

"What's this?"

"A little something to melt your heart."

"It'll take more than a few roses to make up for the years of pain you've caused me."

The wife presumably will not so easily let go of the anger she has nurtured for so long. However, the husband should be prepared for this response and steel himself to counter all her negative responses with positive ones. The session may progress in various ways. "I have more than a few roses," he may say—or, "I know just how angry you are at me, and I'm asking you to just give me a chance this one evening to begin to make up for all the pain. As the Chinese proverb states, "A hike of a thousand steps begins with the first step."

The husband may prepare a candlelight dinner (or even have a catered dinner); give the wife a luxurious bubble bath; sing a special song written for her (or, if he is not up to that, play it as sung by a friend, via tape). The song should be ultra-romantic, extolling the wife's virtues—her beautiful eyes, her winsome smile, her bounteous cooking, etc. However, the coup de grâce will come when he gently nudges her toward the bedroom.

"Where are we going?" the wife may ask, still skeptical.

"You'll see."

"I don't believe you. What are you up to?"

"I'm not up to anything. I've just decided to be nice to you."

He guides her into the bedroom, which has been converted into the bedroom beautiful. It is bathed in hazy dimness, incense is burning, soft music is playing, and one pink light highlights the bed—which is covered with satin sheets. And what is that scattered lavishly on those sheets? Rose petals. Hundreds and hundreds of rose petals. The bed is covered from headboard to footboard with them. They are everywhere! Red, yellow, pink, and white rose petals.

"What's this?" the wife may ask.

"Rose petals for my rose."

"I can't believe you. Where did you get this idea?"

"Oh, from a silly little book I found at the bookstore."

"I want to see that book."

"Later. Let me help you off with your clothes."

"This is too much."

The rose petals usually soften even the angriest heart—and when the heart softens, the body follows. The husband should sieze the opportunity to shower the wife with a very attentive and caring dose of lovemaking, doing the things he knows will make her happy. He can utilize the rose petals, grasping them in his hands and letting them fall on her breasts and belly button. He can do fanciful things, covering her eyes with petals and then kissing the petals. He can also cover parts of his own body with them. (In the event that the wife is so hard-hearted that even the rose petals do not soften her, the husband may use the occasion to discuss what this means about their relationship.)

For most, the lovemaking will be a revelation. After the wife has gotten over her shock, she may be surprised by the kinds of thoughts and feelings—ranging from fear to sadness to envy—that are aroused by the experience. The husband, by letting go of his own resistance, will likewise find that trying to please, rather than distance, his wife arouses new feelings in him. These new thoughts and feelings will lead to a new self-awareness and ability to communicate.

Game 3: How Do I Hate Thee?

Players: Husband and wife.

Activists: Both.

Setting: Home or hotel.

Aim: Put husband and wife in touch with how much they hate each other and why, and encourage them to verbalize that.

Game Plan: This game is inspired by the romantic poem by Elizabeth Barrett Browning that begins, "How do I love thee? Let me count the ways." It was written to her husband, Robert Browning. No couple had a more romantic relationship than they. Kept a virtual prisoner by her father, and too sickly to go out, Elizabeth wrote to Robert, already a famous poet. Their correspondence lasted for years before he helped her to escape from her father's house—and then they were wed.

Elizabeth and Robert were not an angry couple, so they had no need for this game. However, there are many other couples who are and do. Before telling each other how they love each other, these people must reveal how they hate each other. In cases like this, an admission of hate must precede one of love; here, to try to express love without first acknowledging hate invariably results in false love, or mere sentiment. Sentimentalism is a reaction against denied hate.

In this game the husband and wife undress, go to bed, and begin making love. When they are lying in one another's arms, either before or during intercourse, they should take turns telling each other all the ways in which they hate one other. Each should begin with the phrase, "How do I hate thee? Let me count the ways." For example, the wife might say, "I hate you because you never pay attention to me. I hate you because you talk to your mother more than you talk to me. I hate you because you are a selfish bastard." And the husband might say, "I hate you because you're a castrating bitch. I hate you because you pay more attention to our son than you do to me. I hate you because you always get headaches every time I want to have sex with you."

After they have completed this part of the exercise, they take turns saying the same thing to themselves. "How do I hate myself? Let me count the ways." The wife might say, "I hate myself because I'm always so angry and bent out of shape. I hate myself because I'm so oversensitive. I hate myself for rejecting you sexually all the time." And the husband might say, "I hate myself because I can't get an erection. I hate myself because I feel like a failure. I hate myself for being so passive."

The game allows each not only to verbalize the anger that they've been acting out, but also to get in touch with and verbalize the anger at themselves of which they are usually less aware. Having them do the exercise while in the act of lovemaking serves to bring out the erotic elements that attach to the anger, and the sexual atmosphere softens the anger and helps them work through it.

Game 4: Beauty and the Beast

Players: Beauty (wife) and the Beast (husband).
Activists: Both.
Setting: A dwelling in the woods.
Aim: Appeal to husband's angry self and wife's feeling that she is married to a beast—but give her a different way of relating to that beast, designed to transform both him and her.
Game Plan: This game is for angry couples consisting of a husband who has been continually accused of being—and may actually be—a beast, bastard, abuser, or (if possible) worse; and a wife who despises him and despairs of his ever changing. Like other couples, they will have been locked into this ritualized way of relating, she hurling insults at him in an attempt to browbeat or guilt-trip him into being good, and he defying her by being even more beastly.

If possible, the couple should play this game mainly in a house or cabin in the woods. (If they can get hold of a castle, so much the better.) Once in the woods, the game should loosely follow the story line of the famous children's story it is named after. The wife wanders into the woods one day, dressed in a virginal white gown. She encounters the Beast.

"Who are you?" she asks in a frightened tone.

"You can call me Mr. B," the husband growls. He may wear some kind of costume or mask, or he may just use his own sourest expression and most menacing posture. "Come with me."

"Why?"

"Because I made a deal with your father. I paid him a thousand dollars, and now you must be my wife."

"I will not be your wife. You are a beast."

"I said come!"

He "drags" her to the cabin. Once inside it, he asks her to take off all her clothes, and he removes his. He makes himself look as hideous as possible, playing up all the very features that his wife has generally criticized. For instance, if the wife has told him in the past that his belly is big and grotesque, he now protrudes it even more. If she has complained that he is too hairy, he is now twice as hairy (perhaps sporting a beard).

"What do you want with me?" the Beauty asks.

"I told you. To be my wife."

"That's impossible."

"Why?"

"Because you disgust me."

"So?"

"You're a creep."

"So?"

"You're a pig."

"So?"

"You're a beast."

"So?"

"I couldn't stand for you to touch me."

"You'll love me someday."

"Never!"

This conversation (or a variation on it) may be repeated several times over a period of days. At the same time, Beauty repels all of the Beast's advances. However, he takes good care of her in other ways, ingratiating himself to her—preparing meals fit for a princess, entertaining her, tucking her in at night, reading bedtime stories about princes and princesses.

"If you think all this kindness is going to get anywhere with me, you're mistaken," Beauty says. "You still repel me."

"That's all right. I just like being nice to you."

"I don't believe you."

"It's true, my dear Beauty. You will see!"

After days of this treatment, he begins to win her over. Gradually she begins to fall for him.

"You know, you have a nice smile for a beast."

"Thank you."

"And you are a very kind person."

"Thank you."

"Kindness is very important."

"It's nice to have it appreciated."

"Every beast has some good in him, and it's important to cultivate that goodness rather than picking at the badness."

"What a wonderful thing to say. I wish my wife had said things like that."

"You're married?"

"I was. But she left me because she thought I was a hopeless beast."

"Sorry."

Happy ending! They make love and, as in the children's story, the Beast is transformed into a handsome prince. In this scenario, he may actually remove his costume, mask, and wig, or he may simply change his personality. He might also change his clothing, donning a prince's outfit.

"Who are you?" the wife asks.

"You can call me Prince B."

"What happened?"

"Your love for me changed me from a beast to a handsome prince."

"Isn't that amazing! I think I'd like to make love to you again."

"All right—if you insist."

Playing out this game provides an opportunity for the angry husband and wife to step out of their customary mode of relating and see each other from the vantage point of this children's story. Such children's stories are universal because their messages ring true—and, by playing them out, the truths embedded in them (such as that beauty and beastliness are only skin deep) are experienced firsthand by the participants. This game also shows the husband and wife how to convert their anger into sexuality—something they may have forgotten. Finally, it induces them to look objectively at the rut they have been in and the attitudes that have kept them there. (This game can be reversed and played as "Handsome and the Bitch.")

Game 5: Telephone Sex

Players: Wife and obscene phone-caller.
Activist: Husband.
Setting: Telephone.
Aim: To direct anger into an erotic channel not as threatening as direct bodily contact.
Game Plan: This game is designed for angry wives who nevertheless retain a sense of humor.

One evening, when the husband knows the wife is home

alone, he calls her up and surprises her by talking dirty. He may or may not want to disguise his voice at first.

"Hi there, doll. How're you doing?"

"How're you doing yourself. Who's this?"

"This is your mystery caller."

"My mystery caller?"

"That's right. What are you wearing right now?"

"That's none of your business. Is this you, Henry?"

"I told you, this is your mystery caller."

"It's you. I know it's you. What are you up to?"

"I'll bet you have a great body, doll. I think I'd like to see you naked. I think I'd like to run my hand inside your panties. Would you like that?"

"Go to hell."

"I'm getting hard just talking to you."

"Now, cut that out!"

Naturally, the wife will know it's her husband. Thus, if she plays along with the game, it will take off and develop its own plot and dialogue. If she hangs up or becomes upset, the husband may have to try again by seeking her active participation. In either case, the couple should allow themselves to speak to each other as uninhibitedly as possible, calling each other names like "slut" and "creep," thereby giving vent to their anger. By doing so in this unusual way instead of their standard way (which probably consists of arguments in which they arbitrarily tear each other down), they will be seeing themselves in a different perspective. They may even end up talking each other into orgasm, when they have not had orgasms during regular sexual encounters for a long time. Not having to confront one another physically or make eye contact removes some of the risk and resistance alike from the experience. It's like going back to square one.

This game can also be reversed, with the wife playing the role of the obscene phone-caller.

11

Games for Unattracted Couples

Sometimes husbands and wives, as well as couples who have been together for a long time, begin to lose not only their sexual desire but also their sexual attraction. They come to find certain physical features or personality traits unappealing, ugly, or downright repulsive—and they often dwell on these aversive elements and use them as excuses to avoid sex. A bald head or protruding belly or double chin can be the turnoff, or it can be the way a spouse chews gum or laughs or smells. One of my patients had a habit of making a frog sound with her throat each morning before waking up and each night before going to sleep. She had learned this from her mother, who had no doubt learned it from hers. Her husband complained in vain about how unappealing this sound was.

Despite the popular notion that sexual attraction is based mostly on physical attributes, it is in actuality a complex phenomenon made up of many layers. One layer—the most obvious and most conscious one—is the physical appearance of the sexual object. We see a well-proportioned woman in a bikini or muscular man in a loose tunic, and we are immediately attracted. But once a relationship begins (and particularly when it is in its middle stages), physical appearance becomes the least important factor of sexual attraction, while personality and transference become crucial.

Personality becomes crucial because it, more than physical appearance, is what arouses the most intense feelings. Beauty may arouse initial feelings of awe, nervousness, lust, or the like, but personality becomes a permanent factor when a couple settle down into a relationship. The saying "familiarity breeds contempt" seems apt here. What happens with many couples is that certain personality traits eventually are seen as embarrassing, disgusting, or irritating. Such traits become chronic sources of displeasure, disinterest, and distaste.

And personality is linked with transference. As I noted previously, each partner transfers the quality of a relationship from the past with a primary figure—generally a parent or sibling—onto his or her current partner. In cases where one spouse or another (or both) are no longer attracted, the problem is generally associated with a negative transference. That is, the partner begins to remind the other in some way of a repellent brother, sister, mother, or father.

One of my patients, grew up feeling repulsed by his mother's feet, which he said gave off a foul odor. She used to put her feet on his lap while they were watching television, knowing that they would annoy him—then laugh at him when he became angry at her. As an adult, his relationships with women were brief, usually ending after he had sexual intercourse with them. At that point, he would find something about them that disgusted him and quickly reject them. Often some part of their body would disgust him.

The phenomenon of transference and its relation to sexual attractiveness is most clearly demonstrated in the therapy relationship. At some point during therapy, and particularly during psychoanalytic therapy (since in psychoanalytic therapy, the therapy relationship is itself seen as the main agent of change and therefore is intensely analyzed), many patients develop an erotic transference toward their analysts. In other words, the patient falls in love with the therapist. This happens no matter what the therapist looks like—or even smells like. Whatever the case, the patient suddenly becomes enraptured with this counselor, who becomes irresistible—and the patient comes at the analyst with sexual overtures of every variety. Psychoanalysts have discovered that sexual attraction is definitely something that has psychodynamic roots and can be cultivated.

The following games are designed to nourish those roots.

Game 1: Blindfold

Players: Husband and wife.
Activist/s: One or both.
Setting: Home or hotel.

Aim: For those who are visually turned off by their spouse, this game offers a way to eliminate that element and go to a deeper place.

Game Plan: This game is for couples wherein only one partner professes to be unattracted. The other partner surprises the unattracted spouse by putting a blindfold on him or her one evening while they are lying in bed at home, or while they are spending the night at a hotel. "Try it," the activist may say. "This was recommended by a therapist." If the unattracted spouse protests, the other should add, "Try it. What have you got to lose? Maybe you won't be repulsed by me anymore." In cases where both partners are unattracted, this game may be activated by both, with each donning a blindfold. (*Note*: If blindfolds are too uncomfortable, an alternative is to make love in pitch darkness.)

Once the blindfold is (or the blindfolds are) in place, the couple should lie side by side and slowly explore one another's naked body. They should start by touching nonerogenous zones—forearms, elbows, calves, knees, the top of the head. Next they should caress the erogenous zones, but without dwelling on them—the inner thighs, the inner arms, the ears, the back of the neck, the nipples, the lips, the penis, the vagina. Finally, they should make love. No words should be spoken during the game, but the couple are encouraged to make mental notes of their thoughts in order to talk about them later.

The blindfold acts as an artificial barrier, filtering out the visual effects and the negative judgments that inhibit sexual desire. It also serves as an aid to regression: Being blindfolded brings about a feeling of powerlessness and submissiveness associated with childhood, and likewise arouses the primitive erotic feelings known in childhood. Youthful passion, as we all know, is stronger than adult passion mainly because it has not undergone the "thousand and one shocks that the flesh is heir to" (as Shakespeare put it). The game is designed to help participants rediscover both play and the intense feelings of lust they have been inhibiting (and perhaps been afraid of).

At the same time, participants in this game will understand that the physical traits they have been using as excuses to avoid sex are just that. Not that this game will eliminate these excuses or permanently restore sexual attraction. Many (or most) of these excuses to avoid sexual intimacy have transferential roots and hence will be deeply ingrained in the personalities of the participants. They were, after all, developed during childhood to defend against real sexual hurts, and must be respected. However, this game may serve to bring these attitudes to the surface, where they can at last be dealt with.

Game 2: How Do You Repel Me?

Players: Husband and wife.
Activists: Both.
Setting: Home or hotel.
Aim: Get couple in touch with their negative transferences.
Game Plan: This game is a variation of "How Do I Hate Thee?" in the previous chapter. In this version, each spouse verbalizes those things about the other spouse that are sexually repulsive to himself or herself.

As in the previous game, the couple gets undressed, goes to bed, and begins making love. When they are lying in one another's arms, either before or during intercourse, they should take turns telling each other all the things that repel them. The wife might start by saying, "How do you repel me? Let me count the ways." She may then say: "I'm repelled by your bald head . . . by your hairy chest . . . by the fact that you don't bathe regularly . . . by your wimpy smile . . . [etc.]" The husband then takes his turn: "How do you repel me? Let me count the ways. I'm repelled by your fat hips . . . by your constant henpecking . . . by your hairy upper lip . . . [etc.]"

After they have completed this part of the game, they take turns saying the same kinds of things to themselves. "How do I repel myself? Let me count the ways." The wife might say, "I'm repelled by my own body odor, by my own hairy lip, by my squeamish laugh." The husband might say, "I'm repelled by my fat belly, by my passivity, by my hairy chest."

The game allows both partners not only to verbalize the things that repel them, (which they have been acting out by being unattracted to each other), but also the anger at each other for sexually scorning them. (Sometimes couples get locked in a battle of scorn and actually seek to become even more repellent to one another, out of spite.) Each feels rejected by the other and retaliates by counterrejection. Each is also finding distasteful that which they most abhor in themselves (or deny abhorring).

Having the couple do the exercise while in the act of lovemaking serves to bring out the primitive erotic elements that have been submerged by the surface negativity, while at the same time facilitating an awareness of what is being displaced onto the other mate.

For example, they may realize that their objection to their spouse's fat hips has to do with repressed memories about a parent's fat hips, or to childhood taunts about their own fat hips by siblings and schoolmates. Or, they may discover that being repelled by hair is associated with repressed memories about a parent's or sibling's hair. By tracing the transference to its source, the intensity of the transference may be lessened, and sexual feelings enhanced.

Game 3: The Last Person on Earth

Players: Man and woman.
Activists: Both.
Setting: Home, hotel, cabin.
Aim: Stimulate new attitude and new thinking about sexuality.

Game Plan: This game can be played almost anywhere, but it works best in a cabin in the woods, away from other people. The gist of game is that the husband and wife pretend they are living the last day of the world and are the last couple alive. Actually, the game ties in with an unspoken fantasy, as oftentimes unattracted couples will say to each other, "I wouldn't want you even if you were the last person on earth." This remark is more than a statement of fact—it is intended to hurt the other and thereby put an emotional distance between the partners.

The man and woman begin the game by wandering off separately in the woods. They may be wearing ordinary clothes or something special—a ragged and torn shirt, torn jeans, etc. At some point they spot one another and warily make their way toward each other.

"Excuse me," he says. "May I ask if you're the last woman on earth?"

"That's correct. And may I ask if you're the last man on earth?"

"So it seems."

"Then we're stuck with each other."

"I'm afraid so."

"Are you married?"

"I am, but my wife isn't interested in sex."

"Is that so?"

"Yeah. She's always telling me that she wouldn't be interested in me even if I were the last man on earth."

"That's terrible."

"And you? Are you married?"

"Yes, but my husband tells me he wouldn't be interested in me if I were the last woman on earth."

"Very interesting."

"So now here we are, the last man and woman on earth."

"Yes, here we are."

"What shall we do?"

"I don't know."

They can make up their own plot. They may gather wood or go looking for berries, edible mushrooms, and such, or fish in a stream with makeshift poles, or go to the cabin to hide from "danger." It is important that they participate fully in the play-acting of the game, which means getting in touch with the child inside them. Of course, that is half the battle, because it is the fact that they have almost totally squashed their playful inner child that renders them sexually unattracted and uninterested.

Those who cannot playact can still go to a cabin and simply repeat to one another during the course of the weekend, "I wouldn't want you even if you were the last person on earth." Merely repeating this comment again and again will serve to heighten awareness of the intensity of each one's negative transference.

At a certain point the man and woman may go a step further and say:

"I'd rather masturbate than have sex with you."

"Me, too."

They may take turns masturbating, or masturbate at the same time, watching each other and paying attention to the feelings aroused by the experience. Generally, what happens during masturbation is that they find themselves becoming excited by watching the other masturbate. Masturbation is by its nature an exclusionary process, and when we humans feel excluded we become interested in being included. The masturbation may happen several times. Finally, they say to each other (here the husband speaks first):

"Perhaps I could do you by hand."

"That wouldn't be too disgusting?"

"No, I could stand it."

"Then perhaps I could masturbate you as well."

"You wouldn't be repelled by my hairy chest?"

"I wouldn't have to look at it."

"Anyway, we are the last people on earth, so what the hell."

"Yes, what the hell. Go for it!"

They masturbate one another, which generally gets them even more excited. Once they have overcome the initial resistance to making contact, the rest is usually easy. (In the event that any one partner still feels too repelled to continue this part of the game, and says so in no uncertain terms, both should explore why the one feels that way. But then the other partner should say, "Well, let's play the game anyway, and see what happens.") If they soon have become comfortable with the mutual masturbation, they are ready for the next stage. If not, the game can be kept on the verbal level until they are ready.

"Well, being that you are the last person on earth, maybe I ought to just give you a tussle."

"Yes, being that this is the last day in the world, we might as well get our money's worth."

"Even though you're disgusting, I'll try to get through it."

"Even though you're ugly, I'll do my best."

They begin intercourse and continue the dialogue:

"This is disgusting, but I'll press on."

"This is awful, but I'll get through it."

"Your beard scratches me, but it's all right."

"Your breasts are flabby, but I'll hang with it."

"You smell, but I'll live."

"You're too loud, but I'll deal with it."

"After all, you're the last person on earth—so I might as well enjoy it."

"And this may be our last day on earth—so we might as well take advantage of it. One never knows."

These statements have all been said before by these people in one way or another, so it is not cruel to say them in the context of this game (even though it may seem to be). Verbalizing them while having sex and while pretending they are the last two people on earth gives these statements a different meaning than usual. In the past, when such insults were flung at one another regularly, these partners had to develop an emotional numbness or some other defense against them. Now these statements have come alive as though said for the first time. But this time, said with "feeling," they lead to new insights:

"Actually, you might as well be the last person in the world, because I married you and made a vow to you."

"Actually, this could well be the last day on earth, since we never know when we'll die."

"Well, then?"

"Yes, exactly!"

Game 4: Video Sex

Players: Husband and wife.

Activist/s: Either or both.

Setting: Home or hotel.

Aim: To arouse sexual attraction and desire through the use of erotic videos or movies.

Game Plan: If one partner is more unattracted than the other, the other partner may want to activate this game some evening or weekend. He or she may surprise the unattracted spouse by slipping an erotic tape into the video player, then partially or fully disrobing in view of the spouse. Perhaps there will be a seductive wink in the direction of his or her partner.

The activist may choose an erotic video of his or her choice, or one of the classics, such as *Behind the Green Door*, *Misty Beethoven*, *Roommates*, or *Debbie Does Dallas*. After the video has begun to play, he or she masturbates wildly, putting on an act while watching the film, and every now and then glancing at the partner.

"What are you up to?"

"Why don't you join me?"

"No, thanks."

Taking it yet another step, he or she then produces a video camera and asks if the partner would like to join in making an erotic movie.

"Get real."

"I *am*. This is more real than I've been for a long while."

"You're disgusting."

"I know, but why don't you join me, anyway."

"What for?"

"So we can watch a disgusting video or have some disgusting sex."

"No, thanks."

"Try it—what do you have to lose?"

"My sanity."

"Please. I'll pay you five dollars."

"What video did you buy?"

The activist is charming and persistent. Eventually, the resistant spouse joins in. They watch a video or make one themselves. If they watch one, the activist begins to fondle the partner. Invariably such erotic material helps the unattracted spouse to overcome the negative transference.

Next the activist spouse asks the mate to join him or her in acting out the erotic scenes they have just observed in the movie. He or she may say, "Remember when we were kids and we used to act out scenes from the Saturday movies?" In acting out their own erotic scenes, focusing on the re-creation of the scenes from the movie that most stimulated them, they will forget for an hour the blocks that have prevented them from sexual intimacy. The sex may even be surprisingly good (stirred by the erotic video), and afterwards they may even say out loud, "I wonder why I made such a big deal out of your fat hips" or "I can't understand how I used your hairy chest as an excuse to avoid sex for all those years."

As usual, the sex should be followed by a frank discussion.

Game 5: Erotic Barter

Players: Husband and wife.
Activists: Both.
Setting: Home.
Aim: Use bartering to get the uninterested mate to bed.
Game Plan: In this game one spouse attempts to entice the spouse who is the least interested in sex, via the old-fashioned tactic of finding something he or she wants in exchange for sex. For example, let us say it is the wife who is the least interested in sex. Her husband thinks of something she wants from him and offers it as a trade.

"Listen honey," the husband says one Friday evening after dinner. "You know how you're always trying to get me to clean out the attic?"

"Do I? Of course. I've been asking you every day for a year, and you keep saying you'll do it this weekend—but when the weekend comes the attic is still—well—the attic."

"How about if I do it tomorrow morning?"

"Really? I'll believe it when I see it."

"But first you have to do something for me in exchange."

"In exchange? What do you mean?"

"I'm proposing we do a barter. A service for a service."

"What service could I possibly provide you, I wonder."

"Well—funny you should ask. I was kind of thinking about a little sexual service."

"You know I'm not sexually attracted to you."

"I know. I'm not asking you to be attracted. I'm just asking you to lie back and let me do vile and unruly things to your body."

"What kinds of vile and unruly things?"

"Nothing too weird. Just standard stuff. How about it? A clean attic for an hour of indifferent sex."

"And I just have to lie back? I don't have to be interested at all?"

"Right. I'll do everything."

"Do I have to kiss you?"

"Not unless you want to."

"Do I have to hug you?"

"Not unless you want to."

"Do I have to fake an orgasm?"

"Not unless you want to."

"That sounds like a deal I can't refuse."

"I thought so."

"When do we start?"

The husband takes the wife to bed either then or at an arranged time. At first the wife probably will lie back and be uninterested. However, as the husband proceeds to do his "vile and unruly" things to her body, she will undoubtedly have some feelings—whether of sexual arousal or of anger, fear, disgust, and the like. At the very least he will have gotten her past her refusal to have sex. Most likely she will verbalize even more vociferously her objections to other things she finds unattractive about him. He should disregard her comments and continue to enjoy himself. He should say, "I know you don't find me attractive and that's okay. I still enjoy sex with you, and I love you."

This game can be played again and again, using different services as means of barter. As the wife becomes desensitized to her aversion to her husband and more in touch with deeper feelings, their sex life should improve nicely.

12

Games for Politically or Morally Correct Couples

To write a book about sexuality and couples without considering the impact of feminism would be to do it wearing blinders. Feminism arguably has been the most influential cultural movement of the past thirty years in America, affecting all aspects of male–female relations, from dating to sexuality to marriage and childrearing. Unfortunately, although feminism has helped women in many ways, some of its radical notions have been hurtful to marriage.

Most hurtful has been the concretized concept that in relationships between the sexes, men are generally the oppressors or abusers and that women are generally the victims. This concept now permeates even psychotherapeutic literature. A recently published book, *Emotional Abuse*, by Marti Tamm Loring, states in the first paragraph of the preface that masculine pronouns would be used in the book to designate abusers, and female pronouns to designate victims—since in the overwhelming number of cases emotional abusers are men, and in those rare instances where the abuser has been a woman, that was because she was "in defense of being abused by a man." The author then presents a lot of so-called research (mainly case histories by feminist researchers) backing up this claim.

Emotional abuse most assuredly exists—but it is difficult to research, since it is often subtle in nature. Also, it is like a two-way street in that it affords equal passage to both genders. To say that one sex invariably does it in defense of being abused by the other is very much to say the chicken always comes before the egg. (Most wars begin with both sides claiming they

133

are merely defending themselves against the other side's attack.) The breakdown of the family has been facilitated by this radical feminist concept that men are generally guilty and women generally innocent. Please notice that I emphasize that this is a radical feminist concept, not an "all-points" feminist concept. In fact, it is one that many moderate feminists are criticizing.

For example, Christina Hoff Sommers notes, in another recent book (*Who Stole Feminism?*), that radical feminists are so intent on portraying men as abusers that they are willing to bend statistics to make their point. She cites Gloria Steinem's quote in *The Revolution From Within* that "In this country alone . . . about 150,000 females die of anorexia each year." Steinem got this statistic from a book by another feminist who got it from another book by another feminist who noted that eating disorders "are an inevitable consequence of a misogynistic society that demeans women . . . by objectifying their bodies." The third book purported to get its statistics from the American Anorexia and Bulimia Association. However, when Sommers called the AABA, she found that there were 150,000 *sufferers* of anorexia, not 150,000 fatalities. There were actually 54 deaths from anorexia in 1991!

The focus on men as abusers of women and the blaming of men for nearly all of the problems of women is destructive to male–female relations as well as to the family. For if the buck is continually passed to men, then there is no possibility for a real resolution to a marital dispute. Resolutions can be imposed, but they do not last. A real resolution has to do with both sides taking responsibility for their contribution to a conflict, and both recognizing that they are capable of mistakes, excesses, and acting-out. This has been my orientation throughout this book.

Just as these writers misquoted statistics in order to make their point, many women misuse feminism to give themselves a moral advantage over their male partners. It is a way to control their husbands, make them wrong, and avoid real intimacy. Another related problem is that many men—especially those of the passive type—have become so convinced of the rightness of feminism that they dare not do anything to displease their mate, and thus strive always to be politically correct. Even in bed! Often the result is marriages that are "correct," but passionless. If everything must be controlled and correct, spontaneity goes out the window.

Overzealous political correctness is by no means the only excess that can harm a relationship. Religious correctness can also intrude. Some "holier than thou" people will regularly quote from the Bible, or in other ways substantiate or magnify their supposed superior moral correctness, in attempts to control and manipulate their mate. A Scriptures-spouting husband will hammer away at the wife with reminders about how the wife should submit to, and obey, the husband. Such "good advice" only serves to give the husband additional control of the wife—to make of her a kind of "slave of duty" and prevent the development of a voluntary, mutually respectful relationship.

Such modern syndromes have required me to design a special set of games for the "ultra" politically or morally correct couple—that is, for couples in which one or both members misuse political or religious morality. These games, however, will work with any couple using a political posture, religious doctrine, ideology, or philosophy to control sex—an aim which is sure to rob it of spontaneity.

Game 1: Politically Correct Sex

Players: Husband and wife.
Activist: Husband.
Setting: Home.
Aim: In an exaggerated fashion, to mirror the wife's need for politically correct sex so as to point up its absurdity.
Game Plan: One evening or weekend the husband approaches the wife and asks how she would feel about having sex. If she says she would like to, he continues to question her.

"Are you sure you want to have sex?"

"Yes, I'm sure. Why do you ask?"

"Because Robin Morgan, the editor of *Ms.*, says that even when a woman consents to having sex with a man, it's still rape—since all men are in a position of dominance in our society, and hence even when a woman agrees to sex it constitutes giving in to the specter of male dominance and is a kind of psychological rape."

"Yes, I can see the point of that statement."

"So, how do I know you really want to have sex? How do I know you aren't simply giving in to my male dominance now? How do I know it won't be a form of rape?"

"You don't know."

"Then maybe we shouldn't have sex."

"Maybe not. Unless I ask you."

"But even if you ask me, you might be asking me in order to appease a dominant male—and hence it will still be rape."

"True."

"Asking wouldn't be enough. I think you'd have to beg me to have sex, and perhaps take an oath that your desire for sex has nothing to do with wanting to appease me or submit to male domiance but rather has to do with your wish to gratify your own desire."

"That may be right."

"So, start begging."

"Get real!"

This conversation may be repeated numerous times until the wife somehow convinces the husband that she indeed wants to have sex and that the sex is solely to gratify her own desire for him. At that point the actual sexual encounter begins. Throughout this encounter, the husband continues to be exaggeratedly considerate of the wife's feminist stand.

"Would you prefer to undress yourself, or would you like me to undress you?"

"I'll undress myself, thank you."

"Should we lie side by side, so as to be on an equal basis?"

"All right."

"Or would you prefer to get on top, as a kind of affirmative-action sex?"

"Side by side is fine."

"Excuse me. I touched your breast."

"It's all right."

"Are you sure? Are you sure it's not sexual abuse?"

"No, it's not sexual abuse. I want you to touch my breast."

"But are you simply giving in to my male oppression or intimidation?"

"No. I told you, I'm doing it for me. I like doing this, and I'm doing it for me. Okay?"

"You do have nice breasts. Especially your left one. Not that I'm discriminating against your right breast. May I squeeze it?"

"Why must you ask me everything?"

"I don't want to be a sexist."

"You're not being a sexist."

"But I'm using you as a sex object, aren't I?"

"In a way."

"Doesn't that make me a sexist?"

"Not necessarily. Would you just kiss me and shut up?"

"All right. How was that?"

"That was great. Kiss me again."

"Like that?"

"Yes."

"You don't think my tongue is a bit arrogant?"

"Not at all."

"Thanks. Do you mind if I put my hand down there?"

"No, I don't mind."

"You don't think it's rude?"

"No, it's not rude."

"You don't feel sexually exploited?"

"Not really."

"And if I stick my finger inside you, how would you feel about that?"

"Please do."

"You wouldn't feel finger-raped?"

"Just do it."

"How about two fingers? Would two fingers constitute rape?"

"Try three."

"Would you like to get on top now?"

"Why?"

"I'd like to engage in nonsexist intercourse with you."

"Just put it in!"

"But I'm lying on top of you."

"I know. Put it in! Put it in!"

"All right—if you insist. There! Are you all right?"

"Yes, I'm fine."

"Excuse me. I got a little excited. Sorry about that. I lost control of myself for a moment and started spearing you. Did you feel raped just then?"

"No, I didn't feel raped."

"I shouldn't spear, though. I really shouldn't. I'll try to slow down."

"Just do it!"

"But isn't it rude to stick my penis inside you like this? I'm sorry to invade your space like this."

"I want you to."

"You don't feel assaulted?"

"Shut up!"

"You don't feel you're giving in to a male oppressor? You don't feel degraded?"

"Would you stop with the male oppressor stuff?"

"I just want to be politically correct."

"What are you doing? Why are you saying these things? You're mocking me!"

"I just want to make sure everything's politically correct."

"Stop it!"

"But I . . ."

"Stop it, I said, and fuck me!"

Obviously this game involves a kind of teasing by the husband—so it could go awry at any moment. If the wife becomes upset, the husband should drop it and attempt instead to discuss what has happened and what kind of feelings the game has aroused. On the other hand, the wife might find herself becoming both annoyed and passionate at the same time (as in the dialogue above). Her fears of losing control to a man, which she had been warding off through her demands that the sex be politically correct, are now assuaged and made to feel absurd by his ultracautious approach. She suddenly gets in touch with the most primitive well of her sexuality, like the heroines in books like *Lady Chatterley's Lover* (by D. H. Lawrence) and movies like *Swept Away* (by Lina Wertmüller). Down deep she wants to be swept away by passion, as all humans do—but she also fears it. This game helps her to become aware both of that fear and of how it usually is politicized.

Game 2: Morally Correct Sex

Players: Husband and wife.
Activist/s: Wife or both.
Setting: Home or hotel.
Aim: Force husband to confront underlying fears by engaging in a sexual experience that exaggerates moral correctness.
Game Plan: This game is a variation on the previous one, but the goal is the same. The wife can pull this game on her morally self-righteous husband if she thinks that will work, or else elicit his mutual participation in it. In this instance, the wife gently mocks her husband's attitude by being exaggeratedly compliant and religious.

One evening, when the husband comes to her for sex and is about to mount her without any warm-up, the wife stops him.

"Wait, dear. Let's pray first."

"Pray? What for?"

"Because marriage is a sacrament and sexual intercourse is its holy shrine."

"All right—if you insist."

They both kneel down beside the bed.

"Lord, please bless this holy merger of male and female."

"Amen."

"And bless my dearest husband."

"Amen."

"And bless his holy penis."

"Amen."

"Which carries the holy seeds of life."

"Amen."

"And bless this bed upon which we shall lie."

"I think that's enough blessing, dear."

"All right, dear."

They climb back into bed. As the husband begins to mount her again, she speaks again.

"For the Lord sayeth, 'Women, look to thy husbands, and obey them all the days of your lives.'"

"That's right. Just lie back now."

"Anything you say, dear. Are my legs spread out enough?"

"Yes, thanks. They're fine."

"And my vagina—is it wet enough?"

"It could be a little wetter."

"All right. I'll try to make it wetter."

"Thanks."

"And my mouth—is it open wide enough?"

"Do you have to ask me everything?"

"I must look to my husband and obey him."

"You've done enough looking and obeying."

"But the Lord sayeth, 'Cursed be the name Woman.' Didn't He say that, dear? 'Cursed be the name Woman'?"

"He might have. I forget right now."

"Am I cursed, dear?"

"Yes, you're cursed. All women are cursed."

"I'm so sorry I'm cursed. I aplogize for being a woman. Will you forgive me for being cursed?"

"All right."

"Oh, you're sweating. Let me wipe your brow."

"Thanks."

"Am I submissive enough, dear?"

"You're acting like a witch."

"Because the Lord said I'm cursed."

"I suppose so."

"Are you enjoying it, dear?"

"Yes, I'm enjoying it!"

"Anything else I can do?"

"Shut up and fuck me!"

"Oh, dear. What language! After we've just had the Lord bless this sacred bed."

"I said shut up and fuck me!"

If the husband becomes upset, the wife should drop the game and attempt to discuss what has happened and what kind of feelings the game has brought up. Or the husband might find himself becoming more aroused than usual and yelling, "Slut! Witch! Devil!" as he loses complete sexual control of himself. In this case, the discussion should be had as soon as possible after orgasm. The game will help both husband and wife to become aware of how religious excesses have impeded their sexual relationship.

Game 3: Political Reversal

Players: Husband and wife.
Activist: Husband.
Setting: Home or hotel.
Aim: Husband mirrors radical wife's attitude by reversing sexual roles, thereby "shocking" her into awareness.
Game Plan: One evening, when the husband and wife are about to have sex, and the wife as usual begins to lie on top of the husband and take exacting control of their lovemaking, the husbands protests, saying he wants to get on top.

She: "You know I can't stand being on the bottom."

He: "Well, I can't stand it either."

She: "You can't stand it? Why not?"

He: "Because, if you must know, I feel dominated by you. I feel like I'm being raped."

He: "Men can't be raped."

He: "Well, that's how I feel."

She: "Since when?"

He: "Since I had a nightmare the other night."

She: "I don't believe you."

He: "You don't believe I feel raped?"

She: "No."

He: "You see? You don't respect my feelings."

She: "I *do* respect your feelings, but I don't believe what you're saying now. You're just trying to get back at me!"

He: "For what?"

She: "For being so picky about sex."

He: "Do you think you're picky?"

She: "No, but you do."

He: "Do you think I'm being picky now, because I want to be on top?"

She: "I don't believe you mean it."

He: "Suppose I do mean it. Suppose I really do resent having to lie on the bottom all the time? Suppose I really am starting to feel manipulated and raped by you?"

She: "But I'm not manipulating you. I'm just asking you to do something out of consideration for me."

He: "And now I'm asking you to do something out of consideration for me."

She: "But you don't mean that!"

He: "I do mean it! There you go, dismissing my feelings again. I resent that—and I resent your always being on top."

She: "I'm not dismissing your feelings!"

He: "Yes, you are. If you respect my feelings, then you'll give me equal treatment. And equality means that either you let me be on top every other time, or we lie side by side."

She: "I can't believe you're saying this."

He: "Well, I am."

Obviously this conversation can go in many directions. The wife may be amused by it and try to dismiss it as silly. She may become upset. She may simply refuse to have sex altogether. The husband must persevere with this game, however, until she takes him seriously. And when that happens, they will at last have a meaningful discussion about their marriage, their sexuality, and the attitudes that have kept their relationship mechanical and dead.

Various things may come out during these discussions, such as competitive feelings by the wife that include a desire to vilify her husband; and conflicting feelings by the husband—fear, the desire for approval, anger, and the need to defy the wife. The wife may have memories of having been picked on by an older brother, or even molested as a child; the husband may have memories of being his mother's little assistant, or of his father being a womanizer. They should not censor any thoughts that come to them. Saying everything is important, for it unlocks the door within the wall that has kept them rigidly apart sexually.

Continue playing until a sense of real communication and equality is restored.

Game 4: Erotic Cards

Players: Husband and wife.

Activists: Both.

Setting: Home.

Aim: To stimulate a relaxation of counterproductive political correctness by appealing to unconscious competitive feelings.

Game Plan: Most often, a couple whose sex life is constricted by political correctness (or, in the case of religious extremists, moral correctness) is acting out unconscious competitive-ness. This competitiveness is based on narcissism—that is, on the need to bolster lagging self-esteem by defeating the spouse through gaining political and moral superiority. A wife may rule her husband by continually citing rules of feminism that he is breaking, and a husband can rule his wife by continual-ly citing the rules of religion that that she is breaking. Not that there should be no rules or that they should never be cited; but some people, as I have noted, misuse them to put their spouse on the defensive or to gain an advantage that will allow them to defend against their own feelings of low self-esteem and envy.

This game uses the competitive spirit to motivate the polit-ically or religiously correct spouse—whichever that is—to let go of sexual inhibitions that are precluding a more authentic marriage. The couple takes the aces, kings, queens, jacks, and jokers out of an old deck of cards and spreads them on a table—in no particular order. Then they write brief descrip-tions of various kinds of sexual acts on the removed cards, each card citing an act that offends the other person. For instance, on a king they might write, "Let the king take you from behind." On a queen. "Let the queen sit on your throne." On a jack, "Great oral sex." On an ace, "Lie back and submit to heavenly ecstasy." On a joker, "Do with me what you will."

The eighteen cards (sixteen high cards and two jokers) are placed face down on the middle of the table. Then the hus-band and wife take turns drawing one card from the deck. Each time one draws a card, the other asks, "Play or pass?" If the card holder wants to play, he or she must do whatever is written on the card. If he or she says "Pass," that player may keep that card and draw again. Each partner is allowed three passes per draw. However, he or she must put $50 in the jack-pot for *each* pass. After three passes, should a partner refuse to play, he or she loses the game and forfeits the jackpot. (The jackpot includes the sum of money that is begun with—per-haps $100, contributed to equally.) If each performs what the cards dictate, the winner is the one with the fewest passes. If the game is still tied at the end, the jackpot is divided—and each partner becomes a winner in more ways than one.

This game, in effect, puts politically or morally correct

spouses in a no-win situation. If they "play," they lose, since they must submit to some politically or morally incorrect form of sexuality determined by the "opponent." If they refuse, they eventually lose the jackpot. Hence the game tends to bring out feelings that have long been warded off by avoiding certain forms and postures of sexuality. In that sense, the game is also a "no-lose" situation. A tie is the most common ending of this game—for good reason.

Game 5: Dog and Cat Sex

Players: Dog (husband) and cat (wife).
Activists: Both.
Setting: Home.
Aim: Break through inhibitions and political or moral extremes by "stooping" to animal behavior.
Game Plan: Husband and wife don dog and cat masks (respectively) and crouch on the floor on all fours. Otherwise, just like dogs and cats, they are naked. The dog chases the cat around the floor, and the cat scoots away. The dog barks and the cat hisses.

Dog: "Ruff! Ruff! Come here, cat!"
Cat: "Meow! Meow! Not on your life!"
Dog: "I have a little something for you."
Cat: "I'm sure you do."
Dog: "Ruff! Ruff! You're making me mad!"
Cat: "Hiss! Hiss! Stay away or I'll scratch you!"

The dog continues to chase the cat around and sniff at her. He sniffs at her behind. She wiggles it and trots away.

Dog: "Ruff. Ruff! I said come here!"
Cat: "Meow! Meow! Come get me!"
Dog: "If you don't come here, I'll bite you."
Cat: "You'd better not, because I'll scratch you."

Eventually the dog chases down the cat and begins to have sex with her. Like all animal sex, however there is no concern about political correctness. The result, it is hoped, will be true animal passion.

If the couple can get over their inhibitions to playing this game, it can be very enjoyable, liberating, and insight-producing.

13

Games for Abstinent Couples

George and Tina were married for twenty years but never had sex. (They had fumbled with it during their honeymoon, and found that neither was particularly interested in it.) They told friends and relatives that sex wasn't important. What was important was love and companionship. Sex, they maintained, was overrated—blown all out of proportion by the media.

In recent years, more and more therapists, marriage counselors, and sex-therapy clinics have been reporting cases of marital abstinence—cases in which either one or both of the partners claimed to be simply not interested in sex. In some instances those couples (like George and Tina) were never interested. In others, they either gradually or abruptly lost interest. Indeed, lack of interest seems to be the fastest growing sexual malady today.

The causes of sexual disinterest vary. Some individuals have latent homosexual drives that interfere with their sexuality. In fact, this was the case with George and Tina: Both had unconscious homosexual drives that did not come to the surface until they entered couples therapy. Both were very religious also and so could not accept their own homosexual feelings; they thought such feelings were evil and kept them repressed. The result was that they were never really sexually attracted to each other but got married because of their mutually strong religious views, and other factors. After entering couples therapy, at which time they were both in their early forties, they became aware of the conflict between their religious beliefs and their homosexuality and decided to separate and pursue homosexual relationships. Once they had "come out," they became alive again as individuals, whereas before—although they claimed to be quite happy with both their companionship and their religious faith—they were in essence emotionally dead.

As a matter of fact, it was in playing the game "Indecent Proposal" that they first became aware of their built-in conflict. They both realized that they wouldn't be jealous if their spouse had an extramarital heterosexual affair, but would be quite envious about a homosexual affair. Considering why this was so led to a series of shocks and insights. I neither encouraged nor discouraged their homosexuality; however, I recognized that it was the only way in which either might have an authentic sex or love life.

Homosexuality (like other forms of alternate sexual orientation) is rarely changeable. My theories about this are in line with classical psychoanalytical concepts, which hold that sexual orientation is conditioned during the "stage of discovery"—between two and three years of age, which is when children discover the differences in sexual anatomy between boys and girls. How parents respond to the child's sexual discovery and curiosity, what kind of sexuality they themselves model, and how the parents and siblings relate to the particular child, all have an impact on sexual orientation. There may be personality traits that predispose a child toward a homosexual orientation, but I believe that the family environment is the most crucial factor.

I don't try to change a patient's sexual orientation. I believe a person should be encouraged to fulfill the self in whatever way feasible, as long as that doesn't impose on others or break laws. Whatever effect—good or bad—that homosexuality has on society as a whole is something still being debated by social scientists.

Other factors that cause lack of sexual interest are associated with extreme passivity in men and extreme hysteria in women. Some men, due to an accumulation of repressed anger toward their mother, their wives, and women in general, act out that anger by losing interest in sex. Such men usually had mothers who strongly forbade any expression of anger toward themselves, and fathers who were either absent or passive. Since they cannot express the anger or even assert themselves in a way that would indirectly deal with it, they first bottle it up and then act it out in a passive–aggressive way. They are simply not interested in their wives or significant others. They would rather watch a baseball game, work on their cars, go fishing with the guys. Their wives (and women in general) are simply too much trouble to deal with. Sex is a hassle they want to avoid.

Likewise, certain hysterical women go beyond the "Not tonight: I have a headache" stage to a point where they do not want to deal with sex at all. Their sexual drive is diverted to a heightened interest in parenting or into their career. Some women harbor unconscious conflicts about fulfilling their female role of giving birth to and nurturing children. Often they grew up in households in which they witnessed their mother complaining about being trapped in a marriage and unable to fulfill herself. The child grows up wanting to avoid this trap and ends up going to the other extreme and feeling just as miserable as her mother. Such women develop hysterical attitudes toward sex, viewing it only in negative terms.

The games below have been designed to deal with these problems. Hence, they are among the most radical in the book.

Game 1: Sweet Dreams

Players: Husband and wife.
Activist: Either.
Setting: Home or hotel.
Aim: To use dreams as a way to plant seeds of sexual desire, and waking dreams to activate that desire.
Game Plan: This game has two parts. In the first part the husband or wife (whichever is uninterested in sex) makes a concerted effort either to "seed" his or her dreams with erotic contents or merely to become aware of erotic contents that are there but have not previously risen to the surface. The interested spouse can act as a coach during this phase, questioning the uninterested spouse about his or her dreams each morning, and prompting the spouse each night.

"Did you have any sexual dreams last night?" the spouse can ask each morning. "Tell me about them." The telling of the dreams will help the dreamer get in touch with blocked or erotic impulses and help the activist spouse to feel closer. And at night the activist spouse may simply say. "Don't forget to have a sexual dream tonight. And especially try to have one about me, okay? I like it when you get hot for me in your dreams."

These statements by the activist spouse may at first meet with resistance—an angry protest, such as "I'm not intererested in having a sexual dream—especially one about you." The activist spouse may then use the occasion to explore this subject: "Why aren't you interested in having a sexual dream? What's your objection to having a sexual dream? What's your objection to having a dream about me?" This will lead to an exploration of unconscious feelings that are blocking the sex drive. When these are worked through, the uninterested spouse will be able to have a sex dream and report it. This should serve to arouse the latter's latent sexual desire.

Then comes phase two of the game. In this phase the activist, with the permission of the uninterested spouse, wakes the latter up early some morning, during a time when he or she obviously is dreaming. During these early morning hours, pressure caused by sex hormones is at its highest—which is why men generally become erect then. The uninterested spouse will be in a sort of trance state, being still caught up in the twilight sphere of dreams and sleep, and not fully awake. In addition, he or she may well be in the middle of an erotic dream when awakened. Hence the potential for a successful sexual experience will be at its optimum.

"Is this a dream?" the sleeper may murmur upon waking to the feel of the spouse's tongue licking some erogenous zone.

"Yes, dear, a beautiful dream."

"What time is it?"

"Evermore."

"You never kissed me like that before."

"You never dreamed it before."

"Then let me keep dreaming."

"Dream on!"

Game 2: Indecent Proposal

Players: Husband, wife, and "stranger."

"Maybe."

The conversation proceed
sary. When the uninterested
fort, the activists reveal the
a kind of "Candid Camera
range of responses, from sh
dealing with these initial fe
ple alone. They may go in
spend the night.

If the uninterested spo
wife and suitor return to
game there, in which case
bedroom. Here they m
arisen—and, if desired, m

Whether or not they
most important is that
painful and pleasurable
into the meaning of the
communication. It will
had remained unexplor
remained problematic.

Game 3: Shock T

Players: Husband a
Activist: Either.
Setting: Home.
Aim: To shock th
"stupor" of lack of in
Game Plan: Back
according to which s
of hysteria, and a
that, would slap or
or shakee would ex
is reminiscent of su

Let us say that
sits around watchir
or shop. If his wi
grunts a word or
appeases her witl

Activists: Interested spouse and friend.
Setting: Hotel.
Aim: To provoke jealousy and other unconscious feelings that are blocking sexual interest.
Game Plan: It must be stressed that this game is a radical one and should be played *only* if everything else has failed. Basically, in this game the activist spouse gets a friend of the opposite sex to help him or her make the uninterested spouse jealous. The game should be played warily, and the two activists of the game must be prepared to abandon it if it appears too upsetting to the uninterested spouse.

The activist spouse (presented in this scenario as the woman) approaches the sexually uninterested spouse some evening and asks, "Do you remember that movie, *Indecent Proposal?*"

"Yes. What of it?"

"Well, a few days ago I received an indecent proposal of sorts, and I thought I'd better tell you about it."

"I'm listening."

"You remember Martin, the man we met at the club?"

"Yes."

"Well, he's made an indecent proposal."

"What is it?"

"He's offered to pay me—*us*—$1,000 if he could sleep with me."

"Great."

"You're not jealous?"

"Not at all. Anyway, we could use the money."

Of course, the above dialogue could go many different ways. However, in most cases the uninterested spouse will go along with it. Even if that person has jealous feelings, he or she will want to deny them—or they may still be unconscious. The activist spouse will discuss a time and place when the rendezvous will occur—some Saturday night at a local hotel. The activist spouse will meet the suitor that evening and will return the following morning.

"Does that sound all right?"

"Fine."

"Would you like to come along or wait here?"

"Why would I want to come along?"

"Well, Martin offered
us and watch."

"That's very good of
"He really is very cor
"Quite."

"And about the mon
or would a check be all

"Cash would be bett
cent Proposal?"

"I believe so."

The activist spouse
spouse such detailed q
tly tease the uninterest
buried erotic feelings
becomes upset, the g
played only until such
to be uninterested a
anger, or the like.) If
way up to the meeti
actually have the sti
make the proposal di

"Hi, this is Marti
and I were talking t
and ask you about
Proposal?"

The husband m
to the trysting plac
suitor should maint
signs of discomfort
the hotel and Mar

"You know, Be
"Why thank y
"I never notice
like some kind of
"That's a lovel
"And you hav
be caressed."

"Oh, really?"
"And fine, lo
touch."

response. This sort of thing can go on for years—and the wife can become quite frustrated. If she doesn't find an outlet for the anger that accumulates, she will get to where she can no longer even stand to look at him and to where she behaves in exactly the way that is bound to keep him stuck in his uncommunicative place. (Indeed, she generally has an unconscious need to keep him in that place.) He needs to be roused from his stupor and made aware of how he is both suppressing his feelings and precluding not only his sex life but his relationship in general.

So the wife approaches the husband at one of the times he is watching football on TV and says, "You know, dear, I feel like slapping you."

The husband may simply ignore this statement, considering it just another of her complaints. Or he may turn to her and say, "What for?" Or he may even say, "Fuck off."

In any case, the activist replies, "Would you mind if I slap you right now?"

"Of course I'd mind."

"But I'd really like to. I'd really like to—very much."

"Why do you want to slap me?"

"Because I can't stand you. You make me sick when you watch television all the time. The television is more important to you than I am. Much more important."

"Are you going to get on that again?"

"How about if I slap you just once?"

"What good is that going to do?"

"I'm not sure. I'd just like to try it. It might help our relationship."

"How's it going to do that?"

"I can't say for sure. May I slap you just once?"

"I'm watching the football game, for Christ's sake."

"It'll only take a second."

"What's with you, anyway?"

"It's just a slap, that's all. Just a slap between husband and wife. You'll feel a short burning sensation, and your cheek will be red for a while, but there'll be no permanent damage."

"What are you talking about?"

"Please let me. Please."

"No."

"Just once. Just one little slap?"

"I'm trying to watch television!"

"I know. And maybe if you let me slap you, I won't hate you as much, and I won't mind if you watch television anymore."

"You're crazy."

"I know. May I slap you?"

"You'll let me watch television afterward?"

"Yes."

"All right. If you insist. Go ahead—*slap* me."

The wife walks up, pulls her arm back, and slaps the husband with as much of her might as she thinks proper.

"There," he says. "Feel better?"

"Much, much better."

It may not be immediately apparent that anything has happened. The husband may simply go back to his game, and the wife may to her gardening. However, something has happened: She will no longer feel nearly as angry at him, but he will feel more angry at her. As he watches his football game, he will wonder about the slap, and feel resentful. Perhaps later that night, before they go to bed, he will bring it up again. If he doesn't, she should not, either—she should absolutely wait until he does, even it that takes days. If he does not bring it up within a week, she should approach him again, and ask to slap him again. This should be repeated until he does bring it up.

"What was that all about?" he will finally say.

"The slap?"

"Yes."

"I felt like it."

"I know, but why?"

"As I said, I despise it when you ignore me."

"You think slapping me is the answer?"

"You're paying attention to me now, aren't you?"

"Yeah, sure. Because I'm pissed and shocked."

"Fine."

"You want me to be pissed and shocked?"

"Yes."

"I don't understand you."

"Would you like to?"

"I don't know."

"You see? That's the problem. You don't want to understand me—and I don't want to understand you."

This conversation, it is hoped, will continue over a period of days, weeks, or months, until the couple has earnestly begun to discuss their relationship in a fresh way. Perhaps the husband will even say, "Thanks—I needed that."

Game 4: The Experiment

Players: Husband and wife.
Activists: Both.
Setting: Home or hotel.
Aim: To experiment with various forms of foreplay and intercourse so as to find at least one form that excites the uninterested party.

Game Plan: This game is rather simple: Find out what will excite one's uninterested spouse. However, it is precisely the simple thing that often most is overlooked by many couples. They will remain at an impasse for years, each blaming the other for the impasse—but acting in such a way as to maintain it. Often, on an unconscious level, such spouses do not want to know what excites their partner—because they are too angry about things, past and present, to want to give any pleasure to each other.

In this game, the husband and wife sit naked on a bed, facing one another. (Getting to this first step may be difficult in itself, of course, since the uninterested couple will be resistant to doing anything like this. However, it is hoped that one of them, having read this book, will be able to persuade the other by suggesting, "Try it—you might like it.") Once they are seated on the bed, they take turns doing things to try to excite each other. This should be done like a scientific experiment—coolly, without emotion. (The emotion will come later.)

"Does this excite you?" the wife might ask, fondling the man's genitals.

"Not really."

"How about this?" She tickles his testicles.

"Not really."

"How about this?" She tickles his nipples and fondles his penis at the same time.

"Not really."

"How about this?" She kisses him on one of his erogenous zones.

"Not really."

Each time he says "Not really," she must accept that without showing emotion. In other words, she must not take it personally if he is not interested, but instead continue in a neutral way, trying to find the spot or method that excites him. By not taking it personally, she will probably be reversing a trend that has gone on in their marriage for some time. However, by overcoming her old feelings, she will both grow and help him to grow as well.

The impasse that has kept this couple uninterested has to do with either a need to be right, a need for an apology (that is, to be paid for suffering), or a desire for revenge (keeping the other at an emotional distance due to unconscious spite). The typical female spouse resists this game by thinking, "Why should I try to excite or give him pleasure when he never cares about me?" This is the core of the impasse—both spouses waiting for the other to come around. All this must be given up when one deliberately tries to excite the partner.

The wife keeps trying to excite the husband until she finds something that works. Then she stops, smiles, and says, "Later. Your turn."

He proceeds with his part of the experiment.

"Does this excite you?" he asks, sucking her nipples.

"Not really."

"How about this?" Licking someplace else.

"Not really."

"And this?" Blowing in her ear.

"Not really."

He continues until he does find something that works, in the process overcoming his own blocks to wanting to please his wife.

When they have both found something that pleases the other, they begin the next phase—which is to mutually please each other. Again, they should conduct this phase with an attitude of scientific interest. Something may excite them for a moment, then cease to excite them. They must continue to adapt, continue to look for other ways, and never become hurt or annoyed or impatient. Eventually, they will arouse one another to a point where intercourse is possible. After intercourse, they should have a discussion about what they felt and why.

Incidentally: A variation on this game might be called "The Challenge." In this version, the interested spouse challenges the uninterested spouse by saying, "I'll bet I can arouse your sexual interest." The uninterested spouse will of course gladly accept the challenge, seeing it as an opportunity to act out spite. Whether or not the interested spouse can win the challenge, this version can be used as a means of discussing the uninterested spouse's spite and his or her need to resist sexual arousal (seemingly at all costs).

Game 5: Mystery Lover

Players: Uninterested spouse and Mystery Lover (interested spouse).

Activist: Interested spouse.

Setting: Home.

Aim: To arouse the libido of the uninterested spouse by appealing to romantic fantasies.

Game Plan: This game is inspired by the novel and play *Camille.* In that work, the protagonist, a young man, wins the love of a Parisian courtesan by sending her anonymous love letters, flowers, poems, and other gifts. Whereas the book has a tragic ending, this game will have, if played correctly, a happy ending.

The interested spouse plays the part of the anonymous suitor, the Mystery Lover. He (we'll assume in this case it is a he) begins by sending his wife a letter, avowing his love for her. This letter must be sincere, and therefore not overblown.

"Dear Lady, I have been watching you from afar for a long time now. I feel strangely attracted to you, but I'm sad because I feel that your own deepest passion is dormant. I wish I could draw out this passion so that you could be your true, best self. I know that sounds corny, but I really mean it. I don't know the way yet, but maybe I'll find the way. I know that you're married, and I will respect that. Anyway, I'm too shy to reveal myself to you. But know you are loved and admired.—Your Mystery Lover."

This letter should be followed, a week or so later, by a gift—perhaps a dozen roses delivered by hand, with a note:

"Always thinking about you.—Your Mystery Lover."

When the husband sees the flowers, he might ask, "Who are these flowers from?"—to throw her off. Then, a week later, there might be a short poem. It should be simple and not overly sentimental:

"I think of you and wish/That you and I—/I think of you and wonder/What would be if we—/I think of your beautiful, intelligent eyes/And how it would be/if they looked at me/with love/Or surprise./I think of you. . . ."

This poem is followed by little gifts that arrive in the mail, and more flowers—and of course more letters:

"Dear Lady, I wish I could know what you are thinking right now. I wish I could know the passion that lurks inside you, the secret passion that few have known, the passion that yearns for expression. Perhaps this is presumptuous of me, but I do think I know you—know you better than you might imagine anybody could. I know that there's a lot of love inside you, and that that love has often been misunderstood. I think you need somebody who would truly understand you and give you all the space you need to blossom. I think I could be that person, but I don't know if you'd trust me enough to give me the chance. I'm not asking for much at first—just friendship, the chance to prove I care. I do know that if you give me the chance, I'd love to love you as you've never been loved before—with profound understanding of your every sensitivity. Lately, I have had fantasies of revealing myself to you, but I'm still too shy. So I content myself watching you from afar, and with imaginings of you, picturing your thoughtful, wry smile and your eyes looking at me with your typically pointed gaze. . . . Someday!—Your Mystery Lover."

More gifts, more flowers. The letters and poems become more personal, more erotic:

"I must say that I think of kissing you and of how sweet the kisses would taste, and I imagine you becoming wet—not because I'm such a macho man, but because I have won your trust and respect. I see you beaming at me, your intelligent eyes alive again. . . ."

The husband observes his wife throughout all this. When he is pretty sure she has become "hooked" by the Mystery Lover (let's say he catches her looking intently for mail), then he writes a final letter:

"Dear Lady, I believe we are approaching the moment of truth. I don't think I'm going to be able to restrain myself from revealing myself to you. Every day, every hour, I imagine our meeting. I know you'll be gentle—even if you don't like me, even if you don't feel the same way I do. All I ask is that you treat me with respect. I have only tried to love you. If you can't return that love, I'll understand and go away.—Your Mystery Lover."

The moment of truth comes on a particularly appropriate evening. The husband sets the stage while the wife is gone. When she comes home she finds the lights dimmed, her favorite music playing, and a dining table covered with a white tablecloth, with candles, wine, and roses beautifully displayed. Maybe there is a special gift—a bracelet, ring, or necklace. There's also an envelope on her plate that says "To My Dear Lady."

"What's all this?" she asks.

"Read the note," he replies.

She opens the envelope and reads the card in which he confesses the game—and the drama reaches its resolution.

This game has never failed to elicit a response of tenderness, even from the most uninterested spouse. It can, of course, be played by most of the other types of couples described in this book. I have saved it for last because it is the most surefire game in the book—the one to try when all others fail.

14

Games to Restore Tenderness

While the preceding chapters might be called the bread and butter of this book, this chapter might be considered the dessert. Until now, the games have been designed to get couples in touch with the defensive postures and underlying blocked feelings inside them. This constitutes the main work of overcoming blocks to effective relating. Once aspiring lovers have done that, it is relatively easy to restore tenderness. They may not even need to play any of the games in this chapter to do so. However, I include them so as to point out one of the many paths.

Basically, these games are concerned with the expression of feelings connected with tenderness: need, fear, gratitude, respect, trust, love. Of course, these are not the only tender feelings, and there are as many ways to express these as there are couples. This is merely an outline that may or may not apply to each couple. Each will modify the games for their own purposes.

The reason these games may be necessary is that although many couples are able to break the impasse by playing the games previously described, they often end up becoming inundated with new feelings, thoughts, and memories. Some become confused by this inundation and do not know what to do next. Others become exhilarated—but the thrill quickly wears off and is followed by disillusionment when the novelty of the games loses its appeal and they backslide to their former defensive postures. Still others experience what I call "fools' tenderness"—a feeling of joy connected with such newfound sexual satisfaction. This form of tenderness does not come entirely from the heart, but in part also from the unleashed libido, and so is not a grounded kind of tenderness.

Tenderness, according to my definition, is a feeling of agreement and gratitude that results from constructive communication with another being. While sex is a vehicle that may bring people to a state where they have the well-being necessary to communicate in a constructive way, it is only that—a vehicle. It is not the end result. The feelings of agreement and gratitude come about when two people are able successfully to work through blocks and thus make themselves truly vulnerable to one another. Only to the extent that we can make ourselves completely vulnerable to another person and experience the security and comfort of doing so, can we experience the joy of true tenderness. This means being able to risk sharing all our feelings with somebody—beginning with the darkest negatives and ending with the brightest positives—and having that risk pay off.

The games in this chapter will help couples to take the last step toward tenderness and fulfillment. All the games should be played immediately following intercourse, while the couple lies calmly entwined. The dialogues of tenderness may be even more difficult to say than those in any of the previous games: "I love you" is often more difficult to say than "I hate you," although both statements make us vulnerable. When we say "I love you," we are soft and vulnerable, whereas when we say "I hate you," we are hard and vulnerable—that is, more defensive.

At first the players may find that these dialogues arouse feelings other than tender ones. They should go with those feelings, discuss them, work through them—all of that is part of this process. It may even take several practice rounds before the dialogues can be said with anything approaching sincerity.

Game 1: Expressing Need

The husband and wife are lying in a semidark room. They look into each other's eyes (no matter how difficult it is to see or how uncomfortable it makes them) and take turns saying the following sentences:

"I don't want to need you."

"And I don't want to need you."

"If I need you, I'll . . ." (Say whatever comes to mind next.)

"And if I need you, I'll . . ." (Say whatever comes to mind next.)

"My father was so needy. I don't want to be like him."

"My mother was so needy. I don't want to be like her."

"I don't want you to need me."

"And I don't want you to need me."

"If you need me, I'll . . ." (Say whatever comes to mind next.)

"And if you need me, I'll . . ." (Say whatever comes to mind next.)

"My father needed me, and I couldn't stand it."

"My mother needed me, and I hated it."

"I don't want to be dependent on you."

"And I don't want to be dependent on you."

"If I'm dependent on you, I'll . . ." (Say whatever comes to mind next.)

"And if I'm dependent on you, I'll . . ." (Say whatever comes to mind next.)

"My father was dependent on me."

"My mother was dependent on me."

"I don't want to admit I'm needy."

"And I don't want to admit I'm needy."

"To be needy means . . ." (Say whatever comes to mind next.)

"Yes, to be needy means . . ." (Say whatever comes to mind next.)

"I needed my mother and it felt bad."

"I needed my father and it hurt."

"I need you so much and it's all right."

"And I need you so much and it's all right."

"You need me very much and it's okay. I can accept your need and I won't go away."

"And you need me so much and it's all right. I can take it and I won't ever go away."

They should embrace and kiss freely during this game and keep repeating these sentences until they feel connected to them.

Game 2: Expressing Fear

Husband and wife repeat the following sentences:

"I'm afraid to tell you I'm afraid of you."

"And I'm afraid to tell you I'm afraid of you."

"If I admit I'm afraid of you, I'll . . ." (Say whatever comes to mind next.)

"And if I admit I'm afraid of you, I'll . . ." (Say whatever comes to mind next.)

"My father was afraid of my mother."

"My mother was afraid of my father."

"I was afraid of my sister and she picked on me."

"I was afraid of my brother and he took advantage."

"It would disgust me if you were afraid of me."

"It would sadden me if you were afraid of me."

"It disgusted me to see how afraid my mother was of my father."

"It saddened me to see how afraid my father was of my mother."

"I'm afraid of you because . . ." (Say whatever comes to mind next.)

"And I'm afraid of you because . . ." (Say whatever comes to mind next.)

"I'm afraid of myself in relation to you because . . ." (Say whatever comes to mind next.)

"And I'm afraid of myself in relation to you because . . ." (Say whatever comes to mind next.)

"I'm afraid of you and it's all right."

"And I'm afraid of you and it's all right."

"You're afraid of me and that doesn't turn me off."

"And you're afraid of me and it doesn't make me love you less."

"I'm not afraid of what's inside myself, and I'm not afraid of what's inside of you."

"And I'm not afraid of what's inside myself, and I'm not afraid of what you may be feeling about me."

Game 3: Expressing Gratitude

Husband and wife repeat the following sentences:

"I'm afraid to thank you."

"And I'm afraid to thank you."

"If I thank you, you'll . . ." (Say whatever comes to mind next.)

"And if I thank you, you'll . . ." (Say whatever comes to mind next.)

"My mother never thanked my father."

"And my father never thanked my mother."

"My brothers never thanked each other."

"My sisters never thanked each other."

"I don't want you to thank me."

"And I don't want you to thank me."

"If you thank me, I'll . . ." (Say whatever comes to mind next.)

"If you thank me, I'll . . ." (Say whatever comes to mind next.)

"Nobody in my family ever appreciated me, and I don't appreciate you."

"And people in my family didn't appreciate me, and it's hard for me to appreciate you."

"I appreciate you anyway just because . . ." (Say whatever comes to mind next.)

"And I appreciate you just because . . ." (Say whatever comes to mind next.)

"I'm grateful to you and it's all right."

"And I'm grateful to you and it's all right."

"You're grateful to me and it's all right."

"And you're grateful to me and it's all right."

"I'm so grateful to you for being you."

"And I'm so grateful to you for being you."

"You make me grateful to be alive."

"You make me appreciate other people."

"Thank you so much for being kind to me."

"And thank you so much for being kind to me."

"When I think about your kindness, I want to cry."

"And when I think about your kindness, I want to cry."

"If I cry, I'll . . ." (Say whatever comes to mind next.)

"And if I cry, I'll . . ." (Say whatever comes to mind next.)

"I'm so grateful to you and I don't care if it makes me cry."

"And I'm so grateful to you and I don't care if it makes me cry."

"Thank you."

"You're welcome. And thank you."

"You're welcome."

Game 4: Expressing Trust

Husband and wife repeat the following sentences:

"I'm afraid to trust you."

"And I'm afraid to trust you."

"If I say I trust you, I'll . . ." (Say whatever comes to mind next.)

"And if I say I trust you, I'll . . ." (Say whatever comes to mind next.)

"My mother trusted my father, and look what happened."

"My father trusted my mother, and look what happened."

"I don't want you to trust me."

"And I don't want you to trust me."

"If you trust me, I'll . . ." (Say whatever comes to mind next.)

"And if you trust me, I'll . . ." (Say whatever comes to mind next.)

"My mother trusted me too much."

"My father trusted me too much."

"My mother didn't trust me at all.

"My father expected the worst from me."

"I'm afraid to trust you because . . ." (Say whatever comes to mind next.)

"And I'm afraid to trust you because . . ." (Say whatever comes to mind next.)

"I'm afraid for you to trust me because . . ." (Say whatever comes to mind next.)

"And I'm afraid for you to trust me because . . ." (Say whatever comes to mind next.)

"I never trusted anybody in my family."

"And I never trusted anybody in mine."

"I do trust you, and it feels so good."

"And I do trust you, and it feels great."

"It also feels good to be trusted."

"Yes, it does feel great to be trusted."

"If I trust myself, I can trust you."

"And if I trust myself, I can trust you."

"I will be a pleasure to live up to your trust."

"And it will be an honor to live up to your trust."

"Trust me."

"All right. Trust me."

"All right."

Game 5: Expressing Respect

Husband and wife repeat the following sentences:

"It's difficult for me to say that I respect you."

"And it's difficult for me to say that I respect you."

"If I say I respect you, I'll . . ." (Say whatever comes to mind next.)

"And if I say I respect you, I'll . . ." (Say whatever comes to mind next.)

"I feel I'm giving something away if I say I respect you."

"And I feel I'm giving something away if I say I respect you."

"My mother had contempt for my father."

"My father had scorn for my mother."

"My brothers teased my sisters."

"And my sisters degraded my brothers."

"It's difficult for me to accept your respect or admiration."

"And it's very hard for me to accept your respect or admiration."

"If you say you respect me, I'll . . ." (Say whatever comes to mind next.)

"And if you say you respect me, I'll . . ." (Say whatever comes to mind next.)

"I don't feel worthy of your respect because . . ." (Say whatever comes to mind next.)

"And I don't feel worthy of your respect because . . ." (Say whatever comes to mind next.)

"You're not worthy of my respect because . . ." (Say whatever comes to mind next.)

"And you're not worthy of my respect because . . ." (Say whatever comes to mind next.)

"My parents never respected my feelings."

"If I was angry, my parents laughed at me."

"My father laughed at my mother."

"My mother laughed at my father."

"My brothers degraded my sisters."

"My sisters degraded my brothers."

"If you don't respect me, I'll . . ." (Say whatever comes to mind next.)

"And if you don't treat me with respect, I'll . . ." (Say whatever comes to mind next.)

"Actually, I do respect you and it's all right."

"And I respect you, and it's all right."

"You respect me and I deserve it."

"And you respect me and it's fine."

"I respect myself, so I can respect you."

"And I respect myself so I can respect you."

"Every day, in every way, I respect and admire you more."

"And every day, in every way, I respect and admire you more."

"I'll honor you until I die."

"And I'll honor you until I die."

Game 6: Expressing Love

Husband and wife repeat the following sentences:

"I don't feel comfortable saying I love you."

"And I don't feel comfortable saying I love you."

"If I say I love you, I'll . . ." (Say whatever comes to mind next.)

"And if I say I love you, I'll . . ." (Say whatever comes to mind next.)

"People in my family never expressed affection."

"My mother and father never said they loved each other."

"My brothers never acted lovingly to my sisters."

"My sisters never had a kind word for my brothers."

"I don't feel comfortable when you say you love me."

"And I don't feel comfortable when you say you love me."

"If you say you love me, I'll . . ." (Say whatever comes to mind next.)

"And if you tell me how much you love me, I'll . . ." (Say whatever comes to mind next.)

"My mother never said she loved me."

"My father never said he loved me."

"If I say I love you, that might hurt because it will make me realize how much love I haven't had and how much of it I need."

"And if I say I love you, that might hurt because it will make me realize how much love I haven't had and how much of it I need."

"If I feel hurt, I'll cry—and you'll think I'm weak or pity me."

"And if I feel hurt, I'll cry—and you'll think I'm stupid."

"Actually, I do love you—and it feels all right."

"And I do love you—and it's all right to say it."

"You can say you love me, and I can accept it."

"You can say you love me, and it's fine."

"I don't love myself completely, so it's hard for me to accept your love completely."

"And I don't love myself completely, so it's difficult to accept your love completely."

"If I say I love you, I'm stuck with you. What if somebody better comes along?"

"And if I say I love you, I'm stuck with you. What if somebody better comes along?"

"Marriage, as Ingmar Bergman says, is the death of hope."

"Marriage is the birth of contentment."

"I love myself right now, and I love you right now."

"And I love myself right now, and I love you right now."

"Loving you makes me love myself."

"And loving you makes me love myself."

"I'm so glad I can love you and feel loved by you."

"And I'm so glad I can love you and feel loved by you."

"I love you so much, my dearest."

"And I love you so very much, my darling."

"I need so much love."

"And I need so much love, too."

"You are the most important person in my life."

"And you are the most important person in my life."

"I'd do almost anything for you."
"And I'd do almost anything for you."
"I'll always love and honor you."
"And I'll always love and honor you."
"I'll cherish you until I die."
"And I'll cherish you until I die."
Husband and wife kiss softly and tenderly.

15

The Harmonious Couple

 U ntil now I have been describing various games designed to help couples break through the blocks that prevent them from achieving sexual and relational harmony. Now I will define what I mean by harmony. To achieve a goal, one must be able to visualize it.

Unfortunately, most of us grow up in somewhat dysfunctional families. We have rarely (if ever) seen a harmonious marriage, so we have no idea of what one is supposed to be like. Numerous movies, television shows, and books purport to depict healthy relationships, but at bottom each one has its own axe to grind. Professors, politicians, dieticians, and social scientists all have their various slants. Even therapists disagree—depending on which school of therapy they are from. However, there are several points that one can make about healthy relationships with which few would disagree:

1. *Communication:* Good relationships involve constructive rather than destructive communication. Constructive communication leads to closeness. Destructive communication leads to distance. Constructive communication seeks to understand. Destructive communication seeks to explain, deny, rationalize, justify, blame. Constructive communication wants to build. Destructive communication wants to injure.

2. *Trust.* Trust and commitment must be there if any relationship is to work. Building trust is as essential to a harmonious relationship as building chords is to harmonic music. Without formal structure, harmony soon drops to disharmony—and without a structure of trust in a relationship, the relationship becomes disharmonious.

3. *Equality.* Equality means equal respect, not equal roles, income, time, etc. To strive for equality in all areas can lead to a kind of competition and become disastrous for a relationship. However, no matter whether a couple has a one- or two-career marriage, whether the one does housework and the other is a CEO, they can always relate with equal respect with regard to valuing equally each other's feelings, opinions, and roles.

4. *Agreement.* A harmonious relationship is one in which the two partners agree on many—if not most—things. They may both share a cynical view of life (or a religious or political view); they may both have the same taste in art, movies, books, clothes, houses; they may both share a love of the country, sports, music, stamp-collecting. Most important is that they share the same sense of humor.

These four points are of course difficult to achieve even in the best of times. In present-day America, with its emphasis on role equality (an ideal impossible to achieve), instant gratification, and such relational hazards as AIDS, sexual abuse, and entrapment, they are even more so. Perhaps that is why more than 50 percent of our marriages end in divorce, over half of our adult population now chooses to be single, and about a third of all families are single-parent units. In short, for many, the disadvantages of committed relationships and sexuality seem to outweigh the advantages.

In actuality, the advantages of a harmonious relationship with one special person far outweigh the advantages of single-hood. Such a relationship brings out the best in us as nothing else can do. However, few seem capable of achieving a harmonious, truly loving and caring relationship. And if humans come to think that they are not capable of doing something, their tendency is to rationalize that it is not worth doing. Hence many people derogate marriage.

For example, one of my patients stated again and again that she did not want to get married because marriage was a trap. When we analyzed her background, it turned out that she came from an unhappy family in which the mother continually complained about feeling trapped. Another patient came from an abusive family in which the alcoholic parents battered one another with hostile words and sometimes with violence. He was extremely skeptical about the joys of marriage as described on TV shows, and philosophized that marriage simply was not for everybody.

Yet another patient could not separate from his mother—he still lived a block away from her at the age of thirty-seven, had dinner at her house several times a week, and spoke to her almost every night. His mother urged him (almost every week) to get married, and he claimed he wanted to but was invariably critical of all the women he met. Unconsciously, he feared separating from his mother—and she feared his separation from her; her continual prodding of him was a defense against this fear. These and other such patients are among the growing population of singles who do not know, and at least on the surface do not seem to care, what a harmonious relationship is all about.

Cases of Harmonious Relationships

So many people have said to me, "Show me a happy marriage. I've never seen one. I don't even know of any—do you?" They are convinced that happy marriages are a fantasy conjured up by television producers. My reply invariably is, "Yes—in fact I know of several." The following case histories are based on couples I have treated, known, and/or interviewed. They are among the several.

Todd and Charlotte

Todd and Charlotte are entertainers who, unlike most such people, have managed to keep their marriage together for over thirty years. They have separate careers and live apart for months at a time. However, they keep in touch almost daily by phone, discussing their careers, their children, their homes, and whatever else comes to mind. Their communication is constant and constructive, and they have a great deal of mutual respect. Even though his career has been more visible than hers, they value each other equally. "Today my career may shoot up; next year hers may," he says. "What appeals to the public at any particular moment is unpredictable and has nothing to do with my value as a human being or Charlotte's.

Our respect for each other is based on other things." Being entertainers, they also have much in common and are in agreement about most everything—politics, art, and philosophy.

Both claim that their sex life is as good today as it was in the beginning. Even better, because they both appreciate all the years of hard work each has put into the relationship. That aspect, plus their years of other accumulated positive experiences and trust, continues to enrich and intensify their lovemaking in a way that their initial passion did not—and indeed could not.

There were trouble spots along the way, but Todd and Charlotte worked through them rather than give up. For example, several times during their marriage, Todd had brief affairs. While another woman might have sued for divorce, Charlotte stuck with it. "I could always tell when Todd was having an affair," she said. "He wouldn't be able to look me in the eyes when we were having sex, and under questioning he would admit he had slept with a woman. At first I'd be pissed. Then I'd realize that Todd was a handsome man and that women were always coming on to him, and that it was only human for him to give in once in a while. I also knew that he wouldn't have given in to temptation at all if things were good between us. In fact, two of the times he had affairs were when I had just given birth and had stopped paying attention to him. I had become almost totally involved in my child and oblivious to Todd.

"His affairs were attempts to get my attention. When we understood this and discussed it, our relationship went back to normal, and he wasn't tempted again. Actually, each time he had an affair I saw it as a sign that something was wrong between us and as an opportunity to correct it; and each time I discovered that there was something I was doing that contributed to his wandering."

Other trouble spots concerned the children—who, on becoming teenagers, began to have problems. This caused friction in the family in general and between Todd and Charlotte in particular. After their son crashed his second car, they went into family therapy. "I realized that I have this stubborn need to be right all the time," Todd later admitted. "It was getting in the way of my relating to the children. I always want to be seen as Mr. Good Guy, and they didn't necessarily see me that way, but they couldn't tell me that because I wouldn't hear it. So they smashed cars and got into trouble at school." Added Charlotte, "I had the same problem. I was so busy being Mother of the Year that I didn't stop to look and listen to what my kids were trying to tell me until it was almost too late. Actually, it's never too late."

Todd and Charlotte again managed to work through these rough spots (with the help of therapy) and to maintain both their own, and their family's, integrity. They are not the perfect television family who raise perfect children. Rather, they are a hard-working, harmonious couple who have managed to maintain that basic harmony over a period of years by taking responsibility, being willing to look objectively at themselves, and doing whatever was considered necessary to deal with the inevitable crises of their marriage.

Bryan and Sue

Bryan and Sue had been married for forty-one years when he passed away from a heart attack. Each had been married previously. They met when he was thirty-two and she was twenty-eight. They first saw each other at a local German club, and he immediately fell in love with her. According to Sue, she did not fall in love with him immediately but only gradually came to love him. They were married six months after they met. They eventually had one child, a son. Bryan worked for a lumber company all his life, retiring at age sixty-five. Sue was a housewife. Although he was the provider, she was the decision-maker for the household. They both wanted it that way.

Bryan was a happy-go-lucky guy, a good, faithful husband, and a steady, loyal worker. He did not analyze life but did have an instinct for adaptation. He did not vie with others, and nobody vied with him. He was a follower and glad to be one. Bryan adored his wife and seldom even thought about other women. He took his son under his wing and taught the boy everything he knew about sports, carpentry, and the like. He was always whistling and joking, though sometimes his jokes were on the crude side, as when he would break wind and remark, "Paint that one pink." But Sue chuckled at his jokes good-naturedly. Another woman, one not as at home with herself as she, might well have been disdainful of such crudity (reacting narcissistically by viewing it as a derogatory reflection on herself).

Sue did not adore Bryan as he did her, but she did love him dearly. She valued his loyalty and good nature and devilishness, and—yes—even his crudity. They were all part of the man she married. She loved Bryan and she loved their son, Bryan Jr., immensely, doting on him and making him feel special. ("Junior" became a star athelete and later a successful football coach.)

Bryan and Sue had an active life with many friends, were members of bowling and bridge leagues. They owned a cabin by a lake where they would spend summer weekends with friends. Theirs was an ordinary life filled with ordinary pastimes—but it was a contented, productive life: He did his job at the lumber company well; she did her job as housewife well; and they raised a well-adjusted son.

The couple shared an optimistic view of life that helped to buoy and bond them through the years. They also shared a deep commitment to the relationship, thus keeping it alive. They did not have to talk about things very much; over the years they became very familiar with one another and knew or sensed most of what needed to be done and when. They enjoyed a well-tuned, harmonious partnership.

Each day when he came home from work, Bryan would grab Sue's behind as she stood over the stove, pinch her while making a "beep-peep" noise, kiss her, and exclaim. "Gal, I love you as much as the day I met you." And she would give him a mock frown and say, "Stop that, you idiot—I'll spill the peas." At night they would sit cuddled together to watch television. Once a week—sometimes twice—they would have sex. "He's like an ol' puppydog," Sue would say. "And she's like a cat in heat," he would say, winking. Their sex life, like their life in general, was quite ordinary—but its quality was passionate. They did not have high expectations but instead fully enjoyed the simple comfort of their companionship, their son's development, their friends, their hobbies, and their cabin at the lake.

When Bryan died at age seventy-one, more than three hundred people came to his funeral, for he was beloved by relatives, coworkers, and old schoolmates. He was one of those quiet people who live modest, productive lives and thereby subtly enrich the lives of all around them; and she quietly and without fanfare loved him well and made the decisions without lording it over anybody. They were the kind of people you never read about in the newspaper or see on TV news— because they were not superachievers or grandstanders. They just did what had to be done, and quietly did it well.

Sue often thought about Bryan after his death (although fondly rather than sadly), recalling with love even his crudest jokes. In this happy-go-lucky way, he had fulfilled her—and she, accepting him and cherishing him complete with all his flaws plus her own, had fulfilled him.

Tom and Terry

Tom was a bashful billionaire, and Terry was his socialite wife. Their marriage had lasted forty-seven years by the time this book was being written. He had met her when he was thirty and she was twenty. They claimed they both mutually fell in love. When they met he was a brash junior executive who boasted to her that if she married him he would make a million dollars in one year. She said she was not interested in money but rather admired his exuberant, never-say-die attitude. She felt that he was a man who would always respect her, take care of her, and make her feel secure. She turned out to be right.

Theirs was, at least in part, a complementary relationship. He tended to be a workaholic and might have buried himself completely in his work were it not for her socializing influence. He didn't like social gatherings, feeling uncomfortable in any setting other than a business one. She helped him, first of all, to come out of his shell and interact honestly and lovingly with her; and then to learn to tolerate—even to enjoy— social events. On the other hand, she tended to want only to be a social butterfly, and had no understanding of, or appreciation for, such practicalities as paying bills, hiring staff, maintaining the household, and handling crises. His influence helped to balance her excesses.

The crucial element of their successful relationship had to do with their basic acceptance and tolerance of each other's foibles. She did not become upset or feel rejected by his obsessive need to work but understood it as a part of his basic character. "It was part of the whole package," she explains. "I loved the man, and knew that this part of him was something that couldn't be changed, although it could be influenced a little if I did it in a loving way. So I did what I could. Finding creative ways to make money grow was what fueled him, and I knew that and didn't want to interfere with what made him happy. But I also knew I was an important part of his happiness. I used that to get him to moderate himself and become more of a social animal."

For his part, Tom notes: "When I first met her, all she wanted to do was go to parties. I must admit I had some doubts about her, some doubts about her substance. But I was so in love with her that I gave her space. I didn't know why I loved her then, but I realized that it had to do with the fact that she really did have substance—she was strong enough to put up with an absentee husband like me, for instance. And on some level, even though we were so different, we agreed about many things such as child-rearing, politics, philosophy, even agreeing to be different."

Another key to their success was their sex life. Both avow that it has remained vital up the the present:

"When Tom is at the office, he's all business. He's shrewd, stubborn, aggressive. But when he gets into bed with me, the little boy comes out. I mean, he can be strong and dominating in bed, but he can also be—well—infantile. He throws himself into the lovemaking with complete abandonment, the abandonment of a baby. From the moment he starts, he's out of it. There's no self-consciousness there, no intelligent life; he's primitive. His abandonment carries me along with it, and it's great."

Tom adds:

"Terry inspires me. I've never met a woman as inspiring as she is. When I look at her beautiful, calm, confident eyes, my hormones start doing somersaults, and I'm gone. I've never had any sexual problems with her—although I did with women I dated before I met her."

They had two children, a girl and boy, now grown and with families of their own. Along the way there were some dark periods, such as when the daughter went through a time—during her teenage years—of feeling like a second-class citizen because she was a female; and when the son felt Tom was using his wealth to try to control and dominate his life. But throughout such episodes there was a commitment to family communication that kept things from getting out of hand. "We'd have regular family meetings in the den, and we'd all spill our guts until things calmed down," Terry says. "It wasn't easy. In fact, there were days we all wanted to kill each other. But we got through it."

Today the extended family gathers during most big holidays and keeps in touch otherwise throughout the rest of the year.

Al and Betty

Al and Betty met in a 12-step program. They were both recovering from an addiction—she from alcohol, and he from marijuana. Both were also in psychotherapy, where they had spent a lot of time working through feelings connected with having been raised in dysfunctional, sexually and physically abusive families. As such, they had much in common from the start. In the beginning of their sexual relationship, Betty said she had the usual feelings of being sexually exploited. Al reported having the usual feelings of anxiety, which caused him to be impotent. Only after they had been to bed about ten times was Al able to become erect, and only after about twenty times was he able to have an orgasm. "It was like I had to thaw out," he says. "We both had to thaw out and adjust to each other. At first she was just this beautiful female in bed with me. She could have been anybody. I was intimidated by her. Gradually, she became more familiar and more dear, and I relaxed."

They both remember the time when, in the middle of the night, while making love to her, he suddenly went "berserk": "I started kissing her and kissing her and kissing her and I couldn't get in enough kisses; and then I started mumbling, 'I love you, I love you, I love you,' over and over, one hundred times, and kissing her one hundred times all over her face; and it was all like some dream, and I felt as though I had tapped into some repressed, primitive part of myself; and it was great, so spontaneous—it was really me. I'd never experienced anything like it before."

Until then, Betty had continued to feel sexually exploited by, and distrustful of, Al—as she did with all her other men. But she recalls that particular night as the turning point: "When Al went beserk that night I saw him completely differently. I felt the full depth of his loneliness as a human being, the depth of his passion—the passion that had been twisted and muted by his addiction. He was exposing his total need to me, and it made me respond in a like manner." She soon found herself becoming more passionate than ever. That night was the *real* beginning of their fifteen years together.

Al and Betty have now (as of this writing) been sober for sixteen years, and both have been pursuing careers in publishing. Because of their abusive family backgrounds, they chose not to have children. "I was afraid that if I had a kid, I'd abuse it," Betty says. Al adds, "You never know what kinds of feelings will be drawn out of you by children. Unless you're really ready for it, you shouldn't do it." For the most part, their intimacy with one another has sustained them—particularly the sexual bond. There were times when she would get into what she called her "paranoid mood," begin to see him as a "man on the take," and withdraw from him, snap at him, and question everything he did and said. But he was secure enough not to fight back, and the moods passed. During other times, he would get into his attitude that "Women are all teases," and withdraw from her or get angry at her and accuse her of deliberately frustrating him. Fortunately, because both had worked through their transference feelings in therapy, they had strong enough observing egos to understand what was going on and not act out on that too severely. Always, in the end, their sexual bond brought them back in touch with their core feelings, away from their defensive postures.

"There's some kind of primitive bond there that we don't understand, but which keeps us united and contented," Betty says.

"Whatever it is, it works," Al adds. "Don't fix it if it ain't broke, as the saying goes."

If there is one common thread in all of these happy marriages it is the willingness of both marital partners to look at and take responsibility for their own contribution to any conflicts. I have stressed throughout this book, both directly and indirectly through the kinds of games I have devised, how essential it is for both partners to be able to work through defensive postures that prevent resolution. Unfortunately, many people are convinced that their point of view is right; to give up their defensive postures means being willing to acknowledge—at least partially—that they are wrong.

The overall situation is compounded by political or religious movements that reinforce the notion that only one individual in a marriage, by virtue of having the correct political or religious view, is right, and that the other needs constantly to be monitored, saved, redeemed. Radical feminism and fundamentalist religious movements, with their "us against them" attitude, tend to divide relationships rather than unite them. If women are taught by radical feminists (to whom many look for leadership) that all men are oppressors, rapists, wife batterers, and/or child molesters, and that all women's problems result from male oppression, they will not be very much inclined to look objectively at themselves and do the kind of give-and-take working-through that a relationship requires if it is to survive the dark periods that all relationships have.

Sexual Harmony

To be in sexual harmony is to affirm life and negate death. This is the essence of good sexuality, and it takes us back to the dual concept that I introduced at the beginning of this book: Eros and Thanatos. Each time a couple can break through the resistances to effective bonding, they have scored a victory over death. The act itself is a victory, for it affirms their own vitality, arouses their emotions, senses, and intellect, and clears up perceptual confusions. When we are having loving sex, life suddenly takes on a more optimistic meaning and we become more loving and accepting of ourselves and of humanity in general. Our characterological rigidity (a kind of death itself) recedes: We are less obsessive, less impulsive, less hysterical, less critical, less demanding, less opinionated, less argumentative, and generally less intolerant, even—and this is most important—of our enemies.

The act of love is also a victory for Eros over Thanatos

because it is symbolically an act of reproduction—an act that produces new life. The male deposits sperm inside the female, the female drops an egg, and the egg at first resists the sperm (just as the female at first resists the male). An embryo develops into a fetus, which soon becomes ripe and seeks to wend its way out of the womb and into the world. As labor begins, the mother resists the birth of the baby, for it is an event fraught with pain. But, generally, life wins out again. And each overcoming of resistance is a reaffirmation of Eros and a repudiation of Thanatos. Each is also a conquest of pain.

To live is to deal with pain. To die is to give in to it. Hence, the harmonious couple is one that deals successfully with the general pain of existence, as well as the specific pain of their own relationship. There is no such thing as natural harmony. There is only natural chaos. Harmony takes work, and therefore the harmonious couple is one that continually works at it. Chaos happens when people stop working.

All of the foregoing couples worked at their relationships, keeping them in alignment. Their harmony involved day-by-day dedication, commitment, and perseverance. They accepted the total reality of each other as well as of themselves, and did not let their defensive postures run their lives or sabotage their relationships. They remained open and devoted to making the relationship work—no matter the cost in time and effort. They did not let the political trends swirling around them affect their very personal, one-to-one dance of life.

Harmony has been a prime subject of philosophers and psychologists through all recorded time. The following poem by Lao Tzu describes the quality of harmony to which I refer:

> People come.
> People go.
> Three out of ten are always getting born.
> Three out of ten are always getting dead.
> And another three out of ten are trying so hard to live,
> They are dead in the head.
>
> But there is one out of ten,
> So I have heard,
> Who knows how to take life as it turns.

Who can wander the woods
Without fear of tigers or wild bulls;
Who can stroll through an army
Without dread of arrow or spear.
Upon meeting such a person,
A tiger can find no place to drag his claws,
A wild bull no place to sink his horns,
A soldier no place to land his shaft.
If you are at one with life,
You have no place for death.
        —Poem #50 (trans. by Gerald Schoenewolf)

When we achieve harmony with another person sexually, we achieve a moment of immortality. During that moment we are at one with life, and we have "no place for death." During that moment love and life are reborn, and death and destruction—along with all the rigid defensive postures that enact them—are rendered inactive, ceasing to have an influence on our relationships.

References

Freud, Sigmund. *Three Essays on Sexuality*. New York: Basic Books, 1962.

Freud, Sigmund. *Civilization and Its Discontents*. New York: Basic Books, 1962.

Kaplan, Helen Singer. *Disorders of Sexual Desire: And Other New Concepts and Techniques in Sex Therapy*. New York: Simon and Schuster, 1979.

Klein, Melanie. *The Psychoanalysis of Children*, trans. Alix Strachey. New York: Delacorte Press, 1975.

Kohut, Heinz. *The Analysis of the Self*. New York: International Universities Press, 1971.

Loring, Marti Tamm. *Emotional Abuse*. New York: Brunner Mazel, 1994.

Masters, W. H., and Johnson, V. E. *Human Sexual Response*. Boston: Little, Brown, 1966.

Mornel, Pierre. *Passive Men, Wild Women*. New York: Simon and Schuster, 1979.

Schoenewolf, Gerald. *Sexual Animosity Between Men and Women*. Northvale, N. J.: Jason Aronson, 1989.

Sommers, Christine Hoff. *Who Stole Feminism?* New York: Simon and Schuster, 1993.

Spitz, Rene A. *The First Year of Life: A Psychoanalytic Study of Normal and Deviant Development of Object Relations*. New York: International Universities Press, 1965.

Tzu, Lao (500 B.C.), *The Way According to Lao Tzu, Chuang Tzu and Seng T'san*, trans. Schoenewolf. New York: The Living Center Press, 1992.